McF 3

Quantitative ethology
The state space approach

Quantitative ethology
The state space approach

**David McFarland
and Alasdair Houston**

*Animal Behaviour Research Group
University of Oxford*

Pitman Advanced Publishing Program
BOSTON · LONDON · MELBOURNE

PITMAN BOOKS LIMITED
39 Parker Street, London WC2B 5PB

PITMAN PUBLISHING INC
1020 Plain Street, Marshfield, Massachusetts

Associated Companies
Pitman Publishing Pty Ltd, Melbourne
Pitman Publishing New Zealand Ltd, Wellington
Copp Clark Pitman, Toronto

© D. J. McFarland and A. I. Houston, 1981

Library of Congress Cataloging in Publication Data

McFarland, David.
  Quantitative ethology.

  (Pitman international series in neurobiology & behaviour)
  Bibliography: p.
  Includes index.
  1. Animals, Habits and behaviour of—Mathematical models.
  I. Houston, A. (Alasdair).
  II. Title. III. Series.
  QL751.65.M3M35   591.51   81-1246
  ISBN 0-273-08417-8   AACR2

Filmset, printed at The Universities Press, Belfast, N. Ireland
and bound in Great Britain at The Pitman Press, Bath

All rights reserved. No part of this publication may be reproduced, stored in a retrieval system, or transmitted in any form or by any means, electronic mechanical, photocopying, recording and/or otherwise without the prior written permission of the publishers.

Contents

Preface **vii**

1 Behaviour and its causal factors   1
  1.1 Introduction   1
  1.2 Causal factor space   6
  1.3 Motivational isoclines and behavioural tendency   12
  1.4 Summary   16

2 Internal factors in motivation   19
  2.1 The stability of the internal environment   19
  2.2 Behavioural contributions to physiological stability   26
  2.3 The command space and the environment   32
  2.4 Summary   35

3 External factors in motivation   37
  3.1 The significance of environmental cues   37
  3.2 The cue space   39
  3.3 Time cues   44
  3.4 Summary   48

4 The interaction of internal and external factors   49
  4.1 Introduction   49
  4.2 The problem of measurement   49
  4.3 The conjoint measurement approach   52
  4.4 The functional measurement approach   58
  4.5 The problem of dimensionality   60
  4.6 Summary   63
  Appendix 4.1   63
  Appendix 4.2   65
  Appendix 4.3   67

5 Causal models of behaviour sequences   69
  5.1 Classical control systems theory   70
  5.2 A model of the courtship of newts   77
  5.3 The consequences of behaviour   81
  5.4 The concept of observability   83
  5.5 Non-linear systems   85
  5.6 Summary   88

6 Optimal decision-making 89
6.1 Functional aspects of decision-making 90
6.2 Decision-making mechanisms 102
6.3 Summary 108

7 Static optimization 109
7.1 The problem 109
7.2 The optimal solution 113
7.3 Economic parallels 119
7.4 The matching law 124
7.5 Summary 126
Appendix 7.1 126

8 Dynamic optimization 129
8.1 Dynamic programming 130
8.2 Pontryagin's Maximum Principle 132
8.3 Optimal life histories 146
8.4 Summary 148
Appendix 8.1 148
Appendix 8.2 151

9 Goal functions and objective functions 153
9.1 The inverse optimality approach 153
9.2 Decision rules 160
9.3 Adaptation to environmental change 164
9.4 Summary 167

10 Learning and optimization 169
10.1 Preprogrammed learning 170
10.2 Learning in a predictable environment 172
10.3 Learning in an unpredictable environment 180
10.4 Conclusion and summary 185

References and author index 187
Index 203

Preface

Ethology is distinguished from other approaches to the study of behaviour in employing two distinct types of explanation. On the one hand ethologists make use of purely functional explanations derived from evolutionary theory, while, on the other hand, they employ causal or mechanistic explanations derived from some body of behavioural or psychological theory. This duality is characteristic of classical ethology and distinguishes it from other branches of behaviour study. Thus evolutionary biologists, including sociobiologists, offer explanations of behaviour which are based upon the theory of natural selection but require no assumptions about the proximate causes of behaviour, while psychologists and physiologists seek to account for behaviour in terms of proximate causes and mechanisms, but rarely make use of rigorous argument based upon the theory of natural selection.

If classical ethology is characterized by this dual approach to the study of behaviour, then quantitative ethology ought to reflect this duality in formal or mathematical terms. In this book we aim to provide a framework for quantitative ethology which provides a formal link between functional and causal explanations, and which is potentially capable of accounting for the wide variety of phenomena that occur in animal behaviour.

This book is based upon the quantitative ethology course at Oxford University, but it goes beyond the level that an undergraduate student would normally be expected to attain. Our book is primarily intended for those who are already familiar with control systems theory or who have a mathematical background that will enable them to quickly familiarize themselves with this field. This does not mean that the book can be read only by mathematicians. We have attempted to provide a verbal account of our approach, the mathematics being used primarily to justify arguments. The state-space approach, which forms the basis of our theoretical framework, can be understood without recourse to mathematics, and we have made every effort to make ourselves clear to non-mathematicians. In this we have been helped by many of our colleagues in the Animal Behaviour Research Group, who have given us valuable criticism and advice on various parts of the text. We extend our thanks to all who have helped us and hope that they will enjoy reading the book.

Oxford 1981   D.M. & A.H.

1 Behaviour and its causal factors

1.1 Introduction

Traditionally, animal behaviour has been classified into functional categories, such as aggressive, feeding, and parental behaviour. The assumptions have been, not only that the activities within each category subserve a common biological function, but also that they have causal factors in common. For example, the various aspects of feeding and foraging behaviour are said to serve the common function of food intake, and are also thought of as being driven by hunger. A consequence of this attitude has been some tendency for behavioural scientists to specialize in particular subdivisions of behaviour, such as sexual behaviour or sleep, and to study one such 'motivational' system in relative isolation from the others. As the study of animal behaviour becomes more quantitative, however, interactions between motivational systems are taking on considerable importance.

On the basis of the generalized homeostatic type of motivational system illustrated in Fig. 1.1, McFarland (1971) distinguishes between primary and secondary aspects of motivation. This scheme reflects the classical view that physiological imbalances occur both as a result of the action of environmental factors, such as temperature, and as a result of influences from other motivational systems, such as the feeding system. These imbalances are monitored by the central nervous mechanisms, which in turn actuate two types of corrective mechanism. The latter may conveniently be classified into physiological and behavioural mechanisms, which essentially act in parallel to correct the imbalance. An example of a physiological corrective mechanism would be the pituitary-kidney antidiuretic axis. Mechanisms of this type act to conserve the commodity in imbalance, but are not always able to restore the balance. This is the prime function of the behavioural mechanism, the action of which results in intake of the required commodity. Such intake can have three types of effect: (1) It can have purely behavioural consequences, which feed

2 Quantitative ethology: The state space approach

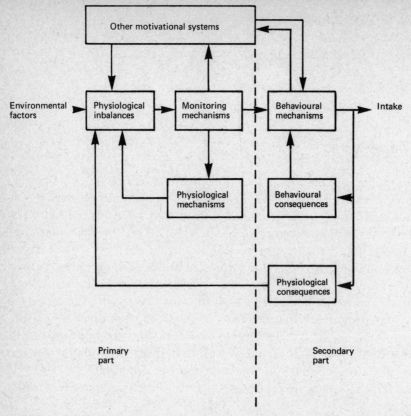

Fig. 1.1 General picture of a homeostatic motivational system. (After McFarland, 1971.)

back to the behavioural mechanism, and subserve satiation, reward, etc. (2) It can have physiological consequences which act to restore the balance. (3) It can influence other motivational systems. For example, ingestion of cold water can have thermoregulatory consequences.

The motivational system portrayed in Fig. 1.1 falls into two distinct parts, separated by a dotted line. The primary part is continuously active and its state at any time represents a situation that can be called the *primary motivational state*. The secondary part is active only when the animal is engaged in the appropriate type of behaviour. The *secondary motivational state* thus refers to those aspects of motivation involved in ongoing behaviour. Interactions between motivational systems can exert their effects on either the primary or the secondary parts. For instance, the state of the feeding system can affect the degree of build-up of motivational potential for drinking, but feeding behaviour itself can also affect drinking directly by its action on the secondary part. Others have also made a distinction between primary drinking and secondary drinking (see Fitzsimons, 1968).

The primary and secondary aspects of motivation can also be differentiated by the methods employed in their study. The typical procedure employed in studies of the primary part is to take food and/or water deprivation time as an independent variable, and to keep physiological and behavioural measures as standardized as possible. Conversely, the secondary part is studied by means of standardized deprivation schedules, whilst manipulating variables directly related to ongoing behaviour during recovery from deprivation such as the availability of food and other environmental factors that affect the consequences of behaviour. The methods employed, and results obtained in experiments of this type have been reviewed extensively by Bolles (1967, 1975).

In terms of the generalized motivational system illustrated in Fig. 1.1, three main levels of interaction may be distinguished. These are summarized in Fig. 1.2. They are the primary level, the secondary level, and the level of the 'final common path' (von Holst and von Saint Paul, 1963; McFarland and Sibly, 1975). At the primary level, the environmental influences may come directly from the environment, as exemplified by the effect of ambient temperature upon the temperature-regulation system, but more commonly they come from the consequences of behaviour (I). Thus water balance may be altered as a consequence of feeding, since many foods contain salts and other substances that obligate water loss (McFarland (1965a)). Moreover, the state of one system may alter the physiological imbalance of another (II). Thus high brain temperature may induce sweating and consequent changes in water balance. At the secondary level of interaction the consequences of one type of behaviour may directly influence the motivational state relevant to another type of behaviour (III). For instance, water ingestion may lead to changes in brain temperature (McFarland and Budgell, 1970), and consequences of reproductive behaviour may lead to shifts in motivational state (Lehrman, 1959; Hinde and Steel, 1966). There may also be direct action of state variables in one motivational system affecting those in another (IV). Thus the evidence suggests that the depressive effect of thirst on feeding is due to inhibition within the central nervous system (McFarland, 1964; Oatley and Tonge, 1969; Rolls and McFarland, 1973).

4 Quantitative ethology: The state space approach

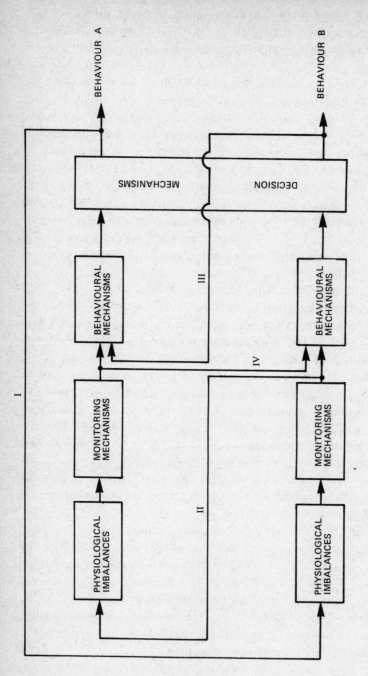

Fig. 1.2 Types of interaction between motivational systems.

Interactions of this type may be shown to be independent of motivational competition at the level of the final common path. For example, Epstein, Fitzsimons and Rolls (1970) observed that a starving rat, which has just been allowed to start eating, changed to vigorous drinking when injected with angiotensin intracranially. To demonstrate that angiotensin genuinely inhibits feeding, it is necessary to show that the effect will occur in the absence of external stimuli associated with drinking, attention to which could compete with those of feeding. This was shown to be the case by McFarland and Rolls (1972).

Behaviour can be classified into various mutually exclusive categories, such that behaviour belonging in one category is incompatible with that belonging in another. For example, if an animal cannot indulge in feeding and sexual behaviour at the same time, it must make a decision between the two activities, and study of this 'final common path' takes us outside the realm of the classical approach to motivation.

One problem is that motivational interactions become so complex, when each is specified separately, that it becomes difficult to maintain a clear picture of the behavioural control system as a whole. Another feature of motivational systems that makes them difficult to envisage as entities separate from each other is that the responses used to assess them are not usually correlated. Hinde (1959) was the first to complain on this account. 'For any pattern of behaviour a number of characteristics may be chosen for measurement, but the correlation between them is often small. The properties of the nervous mechanisms on which they depend thus presumably vary independently, so that the statement that an animal has a strong or weak drive is an inadequate description: use of a simple drive concept thus inevitably leads to an oversimplification of the mechanisms underlying the response'. (Hinde, 1959, p. 130). In an extension of this argument, McFarland and Sibly (1972) propose that drives, or other motivational variables, should be envisaged not as scalars, measurable in terms of magnitude only, but as vector quantities. The consequence of eating, for example, is not simply to reduce hunger, a scalar notion, but to alter many aspects of the animal's internal state, such as salt balance, fat levels, etc. For this reason, it is important to represent hunger as a multidimensional vector quantity, such as that illustrated in Fig. 1.3. McFarland and Sibly (1972) show that the general practice of representing motivational variables as vectors leads to many important concepts that simply do not arise when only scalars are considered; these are discussed in Chapter 2.

6 Quantitative ethology: The state space approach

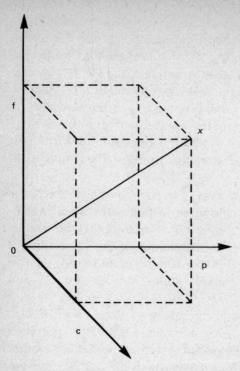

Fig. 1.3 Hunger represented as a multidimensional vector quantity. (After McFarland and Sibly, 1972.) f = fat, p = protein, c = carbohydrate, o = origin, $\mathbf{x}$ = state of hunger.

1.2 Causal factor space

Although McFarland and Sibly (1972) represented the state of a motivational system in a state space, such as that illustrated in Fig. 1.3, they found it convenient to separate the various types of motivation system along traditional lines. More recently, however, McFarland and Sibly (1975) broke with this tradition, and represented the animal's total motivational state in a single *causal factor space*.

McFarland and Sibly (1975) envisage the animal's total behaviour repertoire as controlled by a set of causal factors. These causal factors include variables resulting from the animal's perception of stimuli present in the external environment (e.g. cues to the availability of food), and variables relevant to the animal's internal environment. The state of these causal factors can be represented in a space, the axes of which represent the causal factors. For example, the state might be (hypothetically) represented in a two-dimensional space

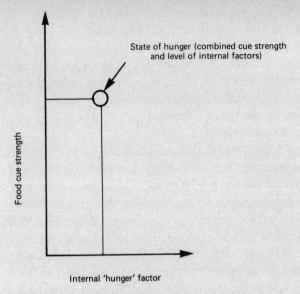

Fig. 1.4 State of hunger in a two-dimensional state space.

(Fig. 1.4) with one axis corresponding to the animal's degree of hunger, and the other to the strength of food cues (i.e. the animal's estimate of the availability of food).

The formulation of the causal factor space depends upon a number of critical assumptions. The first of these (McFarland and Sibly, ASSUMPTION 1) is that it is always possible to classify the behavioural repertoire of a species in such a way that the classes of behaviour, called *activities*, are mutually exclusive in the sense that the members, called *actions*, can only belong to one class. An action is defined as an identifiable pattern of behaviour, such that it is always possible to decide unequivocally whether the action has occurred or not. It is not possible for one action to occur at the same time as another (McFarland and Sibly, DEFINITION 1). In other words, an activity is defined as a set of actions that are mutually exclusive from other activities. This means that no action can belong to more than one activity (McFarland and Sibly, DEFINITION 2).

A second assumption is that the state of the causal factors uniquely determines which action (and therefore which activity) will occur. This means that a particular state of the causal factors will always give rise to the same action, though a particular action might be determinable by more than one state of the causal factors (McFarland and Sibly, ASSUMPTION 2).

8 Quantitative ethology: The state space approach

At any time the causal factors relevant to a number of actions will be present, but by definition only one action can occur. The actions for which the relevant causal factors are present are referred to as *candidates* for the control of ongoing behaviour. By saying that a causal factor is relevant to an activity it is implied that variations in the causal factor will induce changes in the activity (i.e. will lead from one action to another). A causal factor may, of course, be relevant to more than one activity.

Each candidate can be represented along the axis of a space, called the *candidate space*. Each axis of this space represents the strength of candidature, or *tendency*, to perform a particular activity. The candidates can be ordered along each axis in accordance with the following algorithm: In comparison with a particular candidate responsible for activity i, the candidates on any axis j are divided into those that displace i and those that do not. (By displace is meant cause transference of behavioural control from one axis to another, so that a change in activity is observed.) This algorithm applied repeatedly for all candidates on axis i provides an ordering along axis j with respect to axis i. In other words, all possible candidates are sorted with respect to a moving binary criterion on axis i. For example, the algorithm shows that the candidates for two activities can be ordered along the axes of a two-dimensional candidate space, as illustrated in Fig. 1.5. McFarland and Sibly (1975) argue that, not only can the candidates be ordered along the axes of the candidate space, but that the ordering is transitive. Consider a three-dimensional candidate

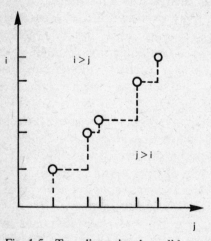

Fig. 1.5 Two-dimensional candidate space, partitioned as a result of the algorithm explained in the text.

space in which particular candidates v_a, v_b and v_c correspond to actions w_a, w_b and w_c respectively. Transitivity means that if candidates v_a, v_b and v_c are ranked such that $v_a > v_b$ so that action w_a is observed in a choice between v_a and v_b; and if $v_b > v_c$ so that action w_b is observed in a choice between v_b and v_c; then $v_a > v_c$ and action w_a will be observed in a choice between v_a and v_c. This argument is based upon the assumption that it is not possible for two candidates to exist such that each displaces the other, without any change in causal factors (McFarland and Sibly, AXIOM 1). Moreover, if v_{a1} and v_{a2} are candidates on the same axis, and v_b is a candidate on another axis such that $v_{a1} > v_b > v_{a2}$, then there exists at least one candidate v_c on a third axis such that $v_{a1} > v_c > v_{a2}$ (McFarland and Sibly, AXIOM 2). These assumptions are sufficient to prove that the ordering of candidates is always transitive. Different approaches to transitivity may be found in Tversky (1969) and Ng (1977). To McFarland and Sibly (1975) transitivity follows naturally from their assumptions about the deterministic nature of behaviour. So a failure to demonstrate transitivity in choice experiments would be regarded by them as a failure to control the conditions of the experiment properly. Other workers prefer to treat transitivity from a stochastic viewpoint (e.g. Navarick and Fantino, 1974).

We can now take the analysis a step further. Consider a two-dimensional candidate space with axes v_A and v_B (Fig. 1.6). In this example a particular candidate always displaces a candidate of lower rank, and equally ranked candidates are impossible (i.e. the shaded set in Fig. 1.6 is a closed set). All cases in which candidates $v_A > v_B$ are shown as shaded area in the figure. The two-dimensional space is divided by a *switching line* on one side of which activity u_A is observed, and on the other side of which activity u_B is observed. In

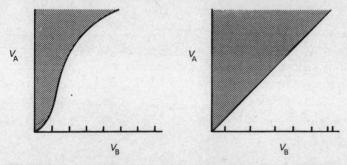

Fig. 1.6 Linearization of the switching line in candidate space by a suitable nonlinear transformation of axis v_B. (After McFarland and Sibly, 1975.)

the limit this line becomes continuous and by means of a suitable nonlinear transformation of one axis (in this case axis v_B) the switching line can be made straight. This example can be generalized to n dimensions (McFarland and Sibly, APPENDIX 2) to show that there is freedom of scaling along only one axis if the switching line is assumed to be straight. Scales along all other axes can then be read off by reference to the switching line. A three-dimensional example is given in Fig. 1.7.

It is important to realize that the candidate space is not a necessary aspect of the causal factor space, but is introduced as a matter of convenience. All that is necessary is that the causal factor space be an Euclidean space with independent axes corresponding to causal factors $x_1, x_2, \ldots, x_n$. For every state of the causal factors, represented by the vector $\mathbf{x}$, there is a corresponding action w_i. The behavioural repertoire of the animal can be conveniently represented in a multidimensional behaviour space, with independent axes $u_1, u_2, \ldots, u_n$ (since the activities are incompatible by definition). Thus there is always some mapping

$$\mathbf{x} \to \mathbf{u} \tag{1.1}$$

and according to the assumptions of McFarland and Sibly (1975) this mapping is unique.

However, the mapping is complex, obscure, and difficult to visualize. The candidate space is introduced as a means of clarifying the picture. As mentioned above, the rationale for the candidate space is

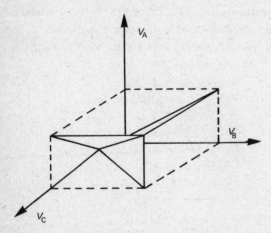

Fig. 1.7 Three-dimensional candidate space showing switching surfaces dividing the space into three parts. (After McFarland and Sibly, 1975.)

the fact that there will always be some activities which the animal will not perform because some of the essential causal factors are missing or very weak. By the same token there will be other activities for which the causal factors are adequate, and which the animal would perform were it not for competition from other potential activities. We are here pinpointing the third level of interaction described by McFarland (1971) and discussed on page 3. The competing potential activities are seen as candidates for control of the behavioural final common path. The strength of the candidature, represented as a point on the axis of the candidate space, is seen as a measure of the animal's *tendency* v_i to perform a particular activity u_i. In animal-behaviour studies it has long been a matter of convenience to specify the underlying tendencies in behaviour, as well as identifying the overt behaviour itself. The question of whether these underlying tendencies influence the overt behaviour is one that we discuss later (Chapter 5).

It follows, from the assumptions and axioms outlined above, that corresponding to every state in causal factor space there can be one and only one candidate state (McFarland and Sibly, THEOREM 1). The candidate state is a point in candidate space, the coordinates being the candidates for each activity.

At this point it may be helpful to introduce a specific, if somewhat hypothetical, example. Zeigler (1974) shows that, when a pigeon feeds from a pile of grain, it takes the grains one at a time with a series of complex and stereotyped movements, which include pecking, mandibulation and swallowing. Pecking consists of a downward movement of the head with the bill initially closed, but gradually opening as it approaches the grain. Contact with the grain terminates this movement, and initiates mandibulation by means of which the grain is propelled from the beak tip to the rear of the buccal cavity. Swallowing is initiated at this point. These three movements occur in a cyclic fashion and are taken by McFarland and Sibly (1975) to constitute an action. To them each action takes about 300 ms, and this duration is not altered by hunger. The increase in overall feeding rate with hunger is brought about by reducing the gaps between each movement-sequence.

We prefer to regard the characteristic pause after each action as part of the action. In this terminology, we suppose that the activity of feeding u_F can take a low value when the food intake rate is low, and a high value when food intake has a high rate. At low feeding rates, the activity will be made up of a series of eating actions, designated

12 Quantitative ethology: The state space approach

w_e, while at higher rates the actions are recognizably shorter, and so may be designated w_f or w_g. Underlying the feeding activity u_F is the feeding tendency v_F, which corresponds to an axis of the candidate space. The particular candidates v_e, v_f, v_g are ranked along this axis and each leads to a corresponding action w_e, w_f or w_g. However, for a particular state of the causal factors x (which may loosely be interpreted as degree of 'hunger' combined with other motivational factors, such as thirst, sleepiness, etc.), there can be only one candidate state v. If this state lies on the feeding side of the switching line in candidate space, then a feeding activity will be observed. If it does not lie on the feeding side of the switching line, then the animal has some other tendency which is stronger than the feeding tendency, so that the transition to overt behaviour $v_F \rightarrow u_F$ will not occur. Although there can be only one candidate for each activity, there can be a set of points in causal factor space all mapping to the same strength of candidature or behavioural tendency. We now consider the implications of this fact.

1.3 Motivational isoclines and behavioural tendency

We have seen that McFarland and Sibly (1975) represent the total motivational state in a causal factor space, in which there is an axis corresponding to each class of causal factor, with the classes defined in terms of some suitable arbitrary criterion. For example, an (hypothetical) animal might have a motivational state that can be represented in a two-dimensional space, with one axis corresponding

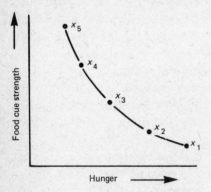

Fig. 1.8 Two-dimensional causal factor space for feeding. The feeding tendency is the same for points $x_1 \ldots x_5$, and the line joining these points is a motivational isocline. (After McFarland and Sibly, 1975.)

to the degree of hunger, and the other to the strength of food cues (i.e. the animal's estimate of the availability of food), as in Fig. 1.4. Thus the causal factor space has axes corresponding to each class of causal factor. It is clear that there are likely to be a number of causal states that will map to the same feeding tendency. For example, a pigeon may have a high hunger but low cue strength (x_1), giving the same feeding tendency u_F as if it had a low hunger and high cue strength (x_5). The line joining all those motivational states ($x_1 \ldots x_5$) which give the same behavioural tendency is an example of a *motivational isocline*, as illustrated in Fig. 1.8.

A simple example of motivational isoclines is provided by the work of Baerends, Brouwer and Waterbolk (1955) on the courtship of the male guppy *Lebistes reticulatus*. The tendency of the male to attack, flee from, and behave sexually towards the female can be gauged from the colour patterns characteristic of each motivational state. In Fig. 1.9 increasing sexual motivation is invoked by colour change and plotted along the abscissa. The effectiveness of the female in eliciting

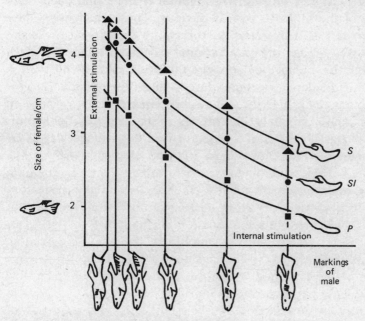

Fig. 1.9 The influence of the strength of external stimulation (measured by the size of the female) and the internal state (measured by the colour pattern of the male) in determining the courtship behaviour of male guppies. Each curve represents the combination of external stimulus and internal state producing posturing (*P*), sigmoid intention movements (*SI*), and the fully developed sigmoid (*S*), respectively. (After Baerends *et al.* 1955.)

courtship increases with her size and is plotted on the ordinate. The points plotted on the graph represent the relationship between the measures of internal state and external stimulation at which particular patterns of behaviour are observed. If the patterns P, Si and S are taken to represent increasing values of response strength, and the scaling of the axes is taken at face value, then the isoclines obtained correspond to those that would result from multiplication of internal and external factors. In this case the method of quantification is somewhat arbitrary, the scaling on the abscissa depending on the association of the different colour patterns with the relative frequency of activities characteristic of sexual tendency and that on the ordinate being arbitrarily linear. Nevertheless, Fig. 1.9 is a good example of the type of representation that we have in mind, and in Chapter 4 we discuss various ways in which arbitrary scaling can be overcome. For our present purposes it is sufficient to note only the principle involved.

As we will see in Chapter 2, the causal factor state is continually changing, partly as a result of environmental changes and partly as a consequence of the animal's own behaviour. These changes in state may be portrayed as a *trajectory* in the causal factor space. As the trajectory moves across the motivational isoclines, there will be a shift in the relative balance of behavioural tendencies. If this shift is such that some tendency becomes larger than the previous largest tendency, a change in activity will occur. Thus every trajectory in causal factor space uniquely determines a sequence of behaviour. However, the characteristics of a trajectory are determined largely by the consequences of the behaviour, but the characteristics of the corresponding behaviour sequence are determined jointly by the path of trajectory and the shape of the isoclines which the trajectory crosses. The trajectories vary from occasion to occasion, but the isoclines remain the same (but see discussion in Chapters 9 and 10). Isocline shape is a design feature, moulded by natural selection, and presumably adapted to the animal's natural way of life. Knowledge of the shape of the isoclines, therefore, is the key to understanding the way in which a particular behavioural system works.

The situation is summarized in Fig. 1.10. For the purposes of illustration, only two dimensions of the causal factor space are shown. All points in plane I map to a single axis in the candidate space (axis v_B), and all points in plane II map to axis v_A. Thus any trajectory in the space (x_1, x_2) will be reflected in corresponding fluctuations in the behavioural tendency v_B. For convenience, the causal factor space is

Behaviour and its causal factors 15

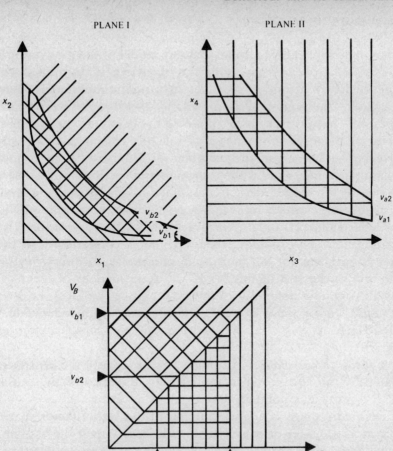

Fig. 1.10 Two planes from causal factor space, and their corresponding regions in candidate space, as explained in the text.

divided up by isoclines which mark the boundaries of states that result in a particular action. These are indicated by the boundaries v_{b1} and v_{b2} in plane I. All points in the space between isoclines map to a corresponding value of the behavioural tendency v_B. The relationship between plane II and axis v_A is similar.

In general, activity u_B will be observed when the behavioural tendency v_b is greater than any other tendency, such as v_a. Thus for particular values of v_A and v_B there will be a point in the candidate space, the candidate state, which will be on one side of the switching line. An example is provided in Chapter 5, and the relationship of tendency to optimality considerations is considered in Chapter 9.

1.4 Summary

McFarland and Sibly (1975) argue that any model of the motivational (i.e. reversible) processes governing the behaviour of an animal can be represented by means of isoclines in a multidimensional causal factor space. The argument is axiomatic, based upon two prime assumptions: That (1) it is always possible to classify the behavioural repertoire of a species in such a way that the classes are mutually exclusive in the sense that the members of different classes cannot occur simultaneously, and (2) these incompatible actions are uniquely determined by a particular set of causal factors. The isoclines join all points in the space which represent a given 'degree of competitiveness' of a particular 'candidate' for overt behavioural expression. The competition between candidates is an inevitable consequence of the fact that animals cannot 'do more than one thing at a time'. In this chapter, and in the rest of this book, we use a number of technical terms which may be defined as follows:

The *causal factor space* is an Euclidean space with independent axes designated $x_1, x_2 \ldots x_n$, and representing the causal factors of behaviour.

The *causal factor state* is a point in the causal factor space, designated by the vector x, and representing the state of the causal factors at a particular point in time.

The *candidate space* is a space related to the causal factor space, formulated as a matter of convenience, and based upon the assumption that some causal factors are so weak that (at a particular time) they have no influence upon behaviour. The candidate space has independent axes $v_A, v_B, \ldots$, etc. corresponding to the strengths of candidature (or tendencies) relevant to particular activities.

The *candidate state*, designated v, is a point in the candidate space corresponding to the strengths of behavioural candidates (or tendencies) at a particular point in time. The candidate state is mathematically the same as the causal factor state, except that less variables are involved in its specification.

An *activity*, designated u_n, is a category of behaviour that is incompatible with other activities.

An *action*, designated w_n, is an identifiable component of an activity.

The *behavioural repertoire* of an animal is ideally made up of a set of mutually exclusive and exhaustive activities.

A *behavioural tendency*, designated v_a, v_b etc., is the strength of candidature for a particular activity, and is represented along an axis of the candidate space.

The *switching line* is the line (or set of surfaces in n dimensions) in candidate space which separates candidate states (combinations of behavioural tendencies) on the basis of their ability to control behaviour.

A *motivational isocline* is a line (or surface) joining all causal factor states that map to the same behavioural tendency. If behavioural tendencies are ranked ordinally, it is convenient to demarcate the boundaries between tendency strengths by means of the motivational isoclines, as illustrated in Fig. 1.10.

2 Internal factors in motivation

2.1 The stability of the internal environment

A general principle that is familiar to most biologists is expressed in Claude Bernard's dictum 'The stability of the internal environment is the condition for free and independent life'. This principle is very general in its applicability, but it has not been exploited to the full by students of behaviour, largely because they have failed to see its relevance. However, every activity affects the stability of the internal environment, because every activity uses energy and involves contributions by various physiological regulatory mechanisms. Moreover, since the performance of one activity may inevitably postpone the performance of other activities, an animal's behaviour may accentuate those aspects of instability that these other activities are designed to minimize. Therefore, the stability of the internal environment must act as a constraint on the way in which motivational systems are designed by natural selection. Consideration of stability is particularly important in the study of interactions between motivational systems, and also in the study of relationships between physiological regulation, acclimatization and behaviour, the three ways by which internal stability is maintained.

The internal environment of an animal can be viewed as a system of interacting variables which is influenced by the animal's behaviour. The state of any biological system can, in principle, be characterized in terms of the state variables of the system, the maximum number of state variables necessary for a complete description of the system being twice the number of degrees of freedom of the system (Milsum, 1966; Rosen, 1970; McFarland, 1971). In other words, the state of the internal environment can be described in terms of a finite number of physiological state variables, each of which is represented as an axis of an n-dimensional hyperspace. Sibly and McFarland (1974) represent the state of the physiological environment as a point in an n-dimensional Euclidean space, the axes of which are independent.

This space is bounded by the physical possibilities (e.g. negative hormonal levels are impossible), and by values of state variables beyond which the animal cannot live (e.g. certain temperature extremes are lethal). For convenience, Sibly and McFarland take as the origin of this physiological space the ideal optimal point on each axis. This point is the value of each state variable that is optimal in the biochemical or physiological sense. It may be that this optimal state is the state that would be attained were external conditions such as to allow the attainment of any of the states within the boundary. This physiological space is illustrated in Fig. 2.1.

The lethal boundary, as conceived by Sibly and McFarland (1974) corresponds to the notion of tolerance limits employed in the study of environmental physiology (e.g. Prosser, 1958, 1973). Adaptation at the tolerance limits, usually called 'resistance adaptations', are greatly influenced by exposure time and by the rate of change of environmental factors. That is, the physiological mechanisms involved in resistance adaptation require a certain amount of time to adjust to a changing situation. If the environmental change is very sudden, death may occur, whereas the animal may be able to survive if the same change takes place more gradually. Sibly and McFarland recognize that the lethal boundaries on different axes will not normally be

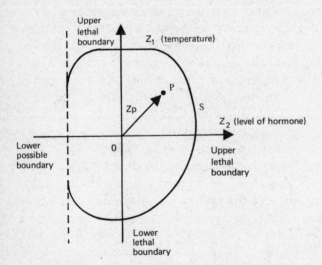

Fig. 2.1 Physiological space. An hypothetical two-dimensional space, with the origin o corresponding to optimal values of body temperature z_1 and hormone level z_2. The physiological state is indicated by the position of the point P, specified by its coordinates, or by the vector z_p. The boundary to the states in which the animal can survive, S, delineates the possible and lethal limits to the values of z_1 and z_2. (After Sibly and McFarland, 1974.)

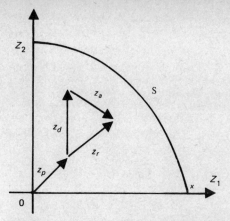

Fig. 2.2 When the physiological state z_p is displaced a distance z_d, physiological adaptive mechanisms act to an extent represented by z_a, so that the resultant displacement is z_r. Other symbols as in Fig. 2.1. (After Sibly and McFarland, 1974.)

independent of each other, and that the exact position of the boundary may depend partly upon the rates of the processes involved. As represented in Fig. 2.1, the physiological state of the animal, denoted by a vector ***P***, is pulled towards the lethal boundary S by environmental forces, which are produced by discrepancies in temperature, salinity (e.g. in a marine animal), etc., between the external environment and the animal's internal environment. When, as illustrated in Fig. 2.2, such environmental influences displace ***P*** by a distance, z_d, various physiological and behavioural adaptive mechanisms come into play, to an extent represented by the vector z_a. If z_a is restricted for some reason so that it is unable to cancel z_d, there will be a resultant displacement z_r.

The adaptive mechanisms or processes which oppose displacement of the physiological state range between rapid physiological reflexes and slow-acting acclimatization mechanisms. There will normally be a spectrum of such processes, as illustrated in Fig. 2.3. Here, for example, a physiological displacement results from the sudden transportation to high altitude, and this is initially counteracted by increased rate of breathing. This type of physiological reflex involves high energy expenditure, and such high cost can be alleviated by means of relatively slow-acting acclimatization mechanisms. An increase in the number of circulating red blood corpuscles is the ultimate response involved in acclimatization to high altitude. Such acclimatization does involve some increased cost, but it alleviates the necessity for extreme behavioural or regulatory measures.

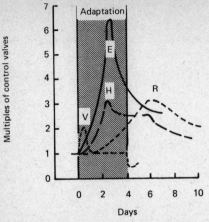

Fig. 2.3 Adaptive modifications in a man breathing rarefied air (p_{O_2}, 85 mm Hg) for 4 days, followed by 6 additional days at sea level. V = lung ventilation; E = serum erythropoietin; H = rate of haemoglobin synthesis, R = fraction of reticulocytes in circulating blood. (After Adolph, 1972.)

In portraying such homeostatic adjustments in terms of the movement of a point in an n-dimensional space, Sibly and McFarland (1974) are simply representing the various processes of homeostasis in multidimensional terms, rather than the more normal unidimensional feedback theory (e.g. Riggs, 1970). McFarland (1971) has already shown that this is an empirically viable approach. For example, the adaptive vector z_a may not necessarily directly oppose the displacement, but may involve further displacement along another axis. This is a fairly common phenomenon in physiological regulation, and is best characterized in a multidimensional space. McFarland and Wright (1969) showed that Barbary doves (*Streptopelia risoria*) conserve water during water deprivation by cutting down on food intake. As a result they become increasingly hungry even though food is available (McFarland, 1964). On the basis of quantitative estimates, McFarland (1971) showed that the deprived state could be represented as a vector in a hunger–thirst state plane. Different experimental methods of estimating the position and movement of the hunger–thirst state (such as estimates from bodyweight measurements, from food–water choice tests, and from amounts consumed during recovery from deprivation) gave good quantitative agreement.

We can distinguish three types of process serving to maintain the physiological state of an animal within lethal boundaries: acclimatization, regulation and behaviour. How do these three processes interact? Sibly and McFarland (1974) maintain that these processes will generally be vectorally additive, and are therefore complementary. In the process of physiological adaptation, various combinations of regulation, behaviour and acclimatization may occur. Thus in adapting to high altitudes, the processes regulating respiration initially function with respect to their normal optima. Stability may be maintained through behavioural measures, such as avoiding heavy work, but gradually the individual will become acclimatized, so that the behavioural measures may be relaxed. At the same time regulation may occur, and be directed towards new optima. Consider another example: a man moving from a cold to a hot climate may be able to expose himself to the sun, acclimatize quickly, and thus obviate any special behavioural measures. On the other hand, he may seek the shade, and thus postpone his acclimatization. These examples emphasize the point that the degree of regulation, acclimatization, or behavioural response depends upon the degree of displacement or drift from the adapted or optimal state. The different processes act in parallel, their effects are additive, but at the same time the success of one mechanism obviates the necessity for another.

Sibly and McFarland (1974) maintain that acclimatization always involves a change in physiological state, and can always be represented by an acclimatization vector. They express the extent to which an animal can adapt physiologically by means of a subspace representing the possibilities of adaptation in terms of limiting velocity vectors. For example, in Fig. 2.4 the rate of drift (unopposed environmental pull) is represented by the length of the vector $\dot{z}_d$. Initially, a strong regulatory response $\dot{z}_p$ may be coupled with a weak degree of acclimatization $\dot{z}_a$, giving a resultant $\dot{z}_r$ sufficient to counteract the drift. At a later stage, however, the contribution from acclimatization may be greater thus permitting a reduction of the regulatory effort. Whatever the combination of mechanisms used in any particular case, the resultant $\dot{z}_r$ must be sufficient to counteract the drift $\dot{z}_d$ if stability is to be maintained. The extent to which the combined adaptation mechanisms are able to operate in any particular case can be represented by an adaptation space Q, defined in terms of limiting velocity vectors (see Fig. 2.5). Sibly and McFarland (1974) claim that any drift vector $\dot{z}_d$ that extends beyond the boundary of the adaptation space Q represents a lethal drift, since the physiological state will eventually drift beyond the lethal boundary S (Sibly and McFarland, THEOREM 1).

This adaptation theorem can be proved under either of two assumptions: (1) That the animal in an inadequate environment adapts

24 Quantitative ethology: The state space approach

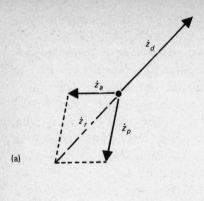

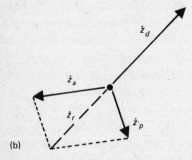

Fig. 2.4 Velocity vector $\dot{z}_d$ representing rate of drift is initially (a) opposed by a strong regulatory vector $\dot{z}_p$ combined with a weak acclimatization vector $\dot{z}_a$, giving a resultant $\dot{z}_r$ sufficient to counteract the drift. At a later stage (b) the contribution of acclimatization may be greater, thus permitting a reduction of regulatory effort. (After Sibly and McFarland, 1974.)

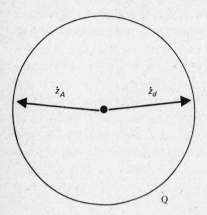

Fig. 2.5 Adaptation space Q. Any rate of drift $\dot{z}_d$ that is greater than $\dot{z}_A$, and therefore extends beyond the boundary Q, represents a lethal drift.

as well as it can, so that the direction of $\dot{z}_A$ is time invariant (Fig. 2.5). (2) That the boundary of the limiting extents to which the adapting mechanisms can act form a convex set. By definition, any two points in a convex set can be joined by a straight line, any point on which is also in the convex set. If neither of these assumptions is valid, then the only strategy which contradicts them but also allows the physiological state to remain within the lethal boundary must employ periodic shifts between one form of adaptation, which counteracts drift in one direction, and another which opposes drift in another direction.

The adaptation vector $\dot{z}_A$ is the sum of the mechanisms serving to counter the imposed drift $\dot{z}_d$. It may be represented as the vector sum of the regulation $\dot{z}_p$, behaviour $\dot{z}_b$ and acclimatization $\dot{z}_a$ vectors, although in reality these processes may not be sharply delineated from each other. However, physiological stability depends upon the efficiency of these processes. In the case of regulation and acclimatization, the efficiency is probably a genetically determined property of the individual. In the case of behaviour, this will sometimes be so but in many instances the animal must learn the appropriate behaviour. Thus the efficiency of the behavioural contribution to adaptation will depend partly upon genetic predisposition towards appropriate behaviour patterns, and partly upon the efficiency of the learning process, which, of course, depends partly upon those features of the environment which provide opportunity for learning. Thus the behavioural contribution to physiological adaptation is likely to be more variable than other regulatory mechanisms.

Consider an animal acclimatized to point z_p in Fig. 2.6. Any drift $\dot{z}_d$ from z_p will act as an input to the acclimatizing systems. Therefore, a necessary condition for the maintenance of the acclimatized

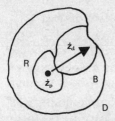

Fig. 2.6 Displacement space D for an animal acclimatized at z_p. When the drift $\dot{z}_d$ remains within D no acclimatization occurs. The contribution of the regulatory mechanisms is indicated by the regulatory space R. The extra contribution of behavioural mechanisms is indicated by the behaviour space B, which can take up any position between D and R. (After Sibly and McFarland, 1974.)

state at z_p, within a period of time t_0 to t_T, is that

$$\int_{t_0}^{t_T} \dot{z}_d \, dt = 0 \tag{2.1}$$

Because of the long time delay involved in the process of acclimatization, small displacements about p will have no effect, provided that their time-average is zero. In situations in which a behavioural response is inadmissible, equation (2.1) depends solely on regulatory mechanisms; here the rate of change of physiological state $\dot{z}_p$ must remain within the *regulatory space* R, illustrated in Fig. 2.6. When relevant behaviour is possible, $\dot{z}_p$ may reach beyond the regulatory space, provided it remains within the larger *displacement space* D, which is the combination of the regulatory space and the *behaviour space* B. The relations between the three spaces R, B and D are illustrated in Fig. 2.6.

Large displacements are often counteracted by behaviour, which returns the physiological $\dot{z}_p$ to the regulatory space. In such cases behaviour is the coarser and regulation the finer mechanism. Clearly, a necessary condition for the maintenance of a particular acclimatized state is that the rate of change of physiological state $\dot{z}_p$ remain within the displacement space D. Provided that the physiological vector $\dot{z}_p$ does not cross the boundary D, the state of acclimatization will remain stable. Similarly, we may visualize a type of regulatory stability characterized by conditions in which regulatory processes alone are sufficient to maintain the state of acclimatization (i.e. $\dot{z}_p$ remains within R). We may regard the boundaries of R as a threshold for the recruitment of behavioural mechanisms.

2.2 Behavioural contributions to physiological stability

We have seen that the processes of acclimatization, physiological regulation and behaviour combine together to form a spectrum of adaptive processes, ranging from slow acclimatization to fast regulation or behaviour. The slow adaptive processes are complementary to the faster mechanisms, in that they obviate the necessity to perform the more expensive short-term activity. Thus the man acclimatized to high altitude (Fig. 2.3) no longer has to breathe so hard. In other words, the effects of behaviour, regulation and acclimatization upon physiological state are vectorially additive (Fig. 2.4).

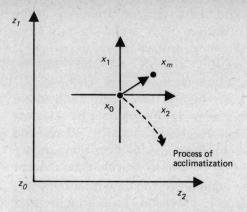

Fig. 2.7 Relation between physiological and motivational space. z_1 and z_2 are the axes of a two-dimensional physiological space, and x_1 and x_2 are the corresponding axes of the motivational space. The acclimatized physiological state x_0 is the origin of the motivational space and the motivational state is x_m. Regulatory aspects are omitted for the sake of simplicity. (After Sibly and McFarland, 1974.)

When vectorially additive processes occur at very different rates, the state of the slow processes can be regarded as the goal of the faster processes (Sibly and McFarland 1974). As depicted in Fig. 2.7, if the animal is acclimatized at state x_0, the regulatory and behavioural processes now aim towards this state, rather than the overall origin z_0. That is, the state x_0 can be regarded as the origin of a 'regulatory' subspace (physiological in nature) and a 'motivational' subspace (behavioural in nature). During the process of acclimatization, the whole subspace moves within the physiological space, as indicated in Fig. 2.7. This representation leads to some interesting predictions about behaviour. For example, as the origin of the subspace shifts in relation to physiological space, the trajectory in physiological space must change as a result of changes in behaviour, if the local origin is to be reached successfully. In other words, the effect of acclimatization upon behaviour is to alter the goal of behaviour. There is some evidence that this does occur. For example, the golden hamster (*Mesocricetus auratus*) prepares for hibernation when the environmental temperature drops below about 15° C. This preparation involves a number of physiological changes amounting to a form of acclimatization to the cold, and makes hibernation physiologically possible when other conditions are favourable. These conditions include the availability of nest material and sufficient food for the animal to set aside a store. Temperature preference tests, conducted in the laboratory, show that the hamsters develop a marked preference for cold environmental temperatures during the pre-hibernating period of acclimatization. They prefer an 8° C environment to 19° C or 24° C. Following arousal from a period of hibernation, the situation is reversed, and the hamsters actively prefer the warmer environments (Gumma, South and Allen, 1967).

Many fish species seem to have highly developed behavioural thermoregulation, and respond to a temperature gradient by selecting a particular temperature (Fry and Hochachka, 1970). Rozin and Mayer (1961) trained individual goldfish to maintain the temperature of their aquarium water by actuating a valve to introduce cold water as the temperature rose. Such fish maintained the aquarium temperature at about 34° C, with considerable precision. To a large degree, the phenomenon of temperature selection explains the distribution of fish in nature (fish generally prefer the temperature to which they are acclimatized (Fry and Hochachka, 1970)). Such an arrangement makes biological sense. If slow changes in physiological state due to acclimatization were not followed by corresponding changes in behavioural goals, then acclimatory processes could be opposed by behavioural mechanisms. For example, acclimatization to cold might be counteracted by a behavioural tendency to select a warmer climate. If a short-lived opportunity to select a warm environment were exploited to the full, then much of the work done in establishing acclimatization to cold would be undone. The obvious alternative is for animals to prefer conditions to which they are acclimatized.

Although the term acclimatization is normally associated with physiological stability, it is convenient to include other slow-acting processes in the same category. Such annual events as hibernation, migration and reproduction involve underlying processes very similar to those that underlie acclimatization. For example, although sexual behaviour does not have consequences which directly affect the physiological stability of the animal, it may be an expensive luxury in that it consumes energy and takes up valuable time that might be more profitably devoted to other behaviour. As McFarland and Nunez (1978) point out, the 'sphere of influence' (displacement space) of restorative processes, such as feeding, thermoregulation, etc., must be sufficient to counteract the physiological consequences of sexual behaviour. When this is not the case, the theory of Sibly and McFarland (1974) is quite explicit as to what will happen. There will be a slow shift of the sphere of influence (see below) to another part of the physiological state space where the consequences of sexual behaviour are not so deleterious or where other aspects of behaviour are better able to contain them. In other words, acclimatization will take place so as to minimize the disruptive consequences of sexual behaviour. In terms of Fig. 2.7, there is a region of the state space where sexual behaviour is 'discouraged', presumably through hormonal influences, and another region where it is facilitated. An annual cycle between these two regions will develop in conjunction with seasonal climatic changes, etc. This type of formulation, therefore, reaches similar conclusions to those traditionally found in the study of long-term behavioural fluctuations but at the same time it provides a means by which such changes can be integrated with other aspects of the animal's behaviour and physiology.

In considering the consequences of behaviour, we can now distinguish the 'goal' (acclimatized state), towards which the regulatory and behavioural mechanisms aim, and the physiological state at any particular time. At any instant, the physiological state will be some distance from the acclimatized state, although this distance will be zero on average, provided the animal remains acclimatized at the same state. It is convenient to take the acclimatized state as the origin of a new sub-space, which may be called a 'regulatory space' where the physiological aspects of homeostasis are the main consideration, or a 'command space' where behaviour is of prime importance. The input to the processes controlling behaviour and physiological regulation is related to the displacement of the physiological state from this local origin, as illustrated in Fig. 2.7. Thus behaviour can often be seen as a parallel activity to regulation, having as its input a displacement in physiological state, and functioning to maintain homeostasis. Many types of behaviour are, of course, not obviously related to homeostasis but this model is a useful one for our present purposes.

The *command space* is so called because displacements of the physiological state from the local origin constitute the actuating variables, or commands, for the mechanisms controlling behaviour. Its axes are obviously related to the axes of the physiological space, but they need not necessarily be the same.

There are many physiological state variables that will not be represented as commands, simply because the animal is not able to respond behaviourally to certain physiological changes. For example, deficiency of vitamins such as thiamine does not lead directly to specific vitamin-seeking behaviour, although it may contribute to the animal's general strategy in coping with sickness (Rozin and Kalat, 1970; McFarland, 1973). Sibly and McFarland (1974) argue that the axes of the command space must be in a one-to-one relation to the axes by which the consequences of behaviour can be represented. Since the consequences of behaviour are determined partly by the nature of the environment, the axes of the command space cannot be a fixed property of the animal. McFarland and Sibly (1972) distinguish between the 'ideal command space' and the 'minimal command space'. The *ideal command space* is a vector space of finite dimensionality, in terms of which every possible command vector can be represented. It is regarded as a fixed property of the individual animal, representing the animal's potential to generate commands relevant to its overall behavioural repertoire. As Richter (1943) showed, when physiological regulators were surgically eliminated '...the animals themselves made an effort to maintain a constant environment or homeostasis'.

The *minimal command space* is the space of minimum dimensionality in which it is possible to represent the consequences of the animal's behaviour. McFarland and Sibly (1972) claim that, provided certain circumstances are met, it is always possible to represent an animal's command state in a natural space of known dimensions (i.e. the minimal command space). A mathematical proof of this claim amounts to the following: provided that the animal is adapted to its environment, and the structure of the environment is known, the command state can be represented in a space, the axes of which can be specified directly in relation to the consequences of behaviour. By 'structure of the environment is known' it is meant that the relationship between the behaviour and its consequences is known. In general, McFarland and Sibly (1972) envisage that once an animal has become physiologically acclimatized to a particular environment it can be said to have reached a state of equilibrium, involving compromise between various physiological systems. These physiological adjustments will generally result in alterations at the motivational level. For example, seasonal climatic changes can induce alterations in hormonal state that affect sexual behaviour. Rationing of food or water induces change in the hunger and thirst and thermoregulatory systems, which appear to result in altered commands to the respective behavioural control mechanisms. Moreover, the consequences of behaviour may be such that ways in which the animal can change its command state are constrained (see below). The command space characteristic of the acclimatized animal is stable and intimately related to the state of the environment. To take a hypothetical example, let us suppose that the ideal command space had coordinates x_c (carbohydrate), x_p (protein) and x_f (fat). The animal is maintained on a fixed diet composed of red food and blue food, in which the red food is always composed of 100% fat, and the blue food is 60% protein and 40% carbohydrate. Because protein and carbohydrate are always coupled together, movements in the ideal command space resulting from the consequences of feeding will always be restricted to a particular plane, as illustrated in Fig. 2.8. Moreover, if the animal is able to adapt to the diet, in the sense that it can survive on red and blue food alone, then movements in the ideal command space which are induced by deprivation will normally occur in the same plane. In other words, in the adapted animal the dimensionality of the motivational space has been reduced from three to two.

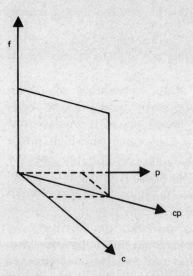

Fig. 2.8 Reduction of a three-dimensional (ideal) command space to a plane by coupling carbohydrate c and protein p in the diet (After Sibly and McFarland, 1974.) f = fat.

The general conclusion that the axes of the minimal command space can be constructed, provided that the structure of the environment is known, is of some practical importance because it offers a way of representing the results of laboratory experiments designed to measure changes in motivational state. It is also important in any detailed representation of causal factor space (see Sibly and McFarland, 1974).

2.3 The command space and the environment

In the type of motivational analysis outlined here, consideration of the effects of environmental changes are important in three different ways.

(1) Changes in the environment may directly influence the animal's physiological state, and thus indirectly alter its command state. For example, a sharp drop in environmental temperature may induce increased heat loss, a fall in body temperature, and an increased motivation for thermoregulatory behaviour. This type of environmental influence is dealt with in the earlier parts of this chapter.

(2) Changes in the environment may alter the relationship between behaviour and its consequences. We follow McFarland and Sibly (1972) in representing this relationship by the environmental matrix E, viz.

$$\dot{c} = E u \qquad (2.2)$$

where $\dot{c}$ is the change in command state consequent upon the behaviour u. Although behavioural consequences have many aspects, affecting orientation and other dynamic properties of the behaviour, we are concerned here only with those consequences which relate directly to the command state. As we have seen, in the case of feeding the consequences might (hypothetically) be formulated in terms of the amounts of carbohydrate, protein and fat obtained during a specified period of time.

(3) Changes in the environment may provide stimuli which the animal evaluates and which induce changes in motivational state. Strictly, both changes in command state and motivational changes due to environmental stimuli are represented as changes in causal factor space (Chapter 1), but it is sometimes convenient to treat them separately. Accordingly, discussions about changes in command state (e.g. McFarland and Sibly, 1972; Sibly and McFarland, 1974; McFarland, 1978a) are based on the assumption that the animal is in a constant environment as far as environmental stimuli are concerned. Consideration of the changes in motivational state that result from external stimuli is the subject of Chapter 3. Meanwhile, the present discussion takes the command state as a subset of both the physiological space and the causal factor space, and assumes that environmental stimuli remain unchanged during the period of observation or experiment.

Suppose a hungry animal were presented with red and green food to which it is acclimatized. Let the constitution of these foods be represented by an environmental matrix E, such that

$$x_F = \int_0^t E u \, dt \tag{2.3}$$

where u is specified in terms of rate of consumption of red and green food, and x_F represents the obtained quantities of protein and fat, corresponding to the axes of the command space. Thus

$$x_F = \begin{bmatrix} p \\ f \end{bmatrix} \tag{2.4}$$

The consequences of ingesting red and green food can be represented in a consequence space of two dimensions (Fig. 2.9) corresponding to the components of the x_F vector.

Suppose the constitution of the green food were 100% protein and that of the red food were 100% fat. Then the environmental matrix would be as follows:

$$E = \begin{bmatrix} 0 & 1 \\ 1 & 0 \end{bmatrix} \tag{2.5}$$

McFarland and Sibly (1972) use the term *univalent* to describe such an environment, in which there is a one-to-one relationship between the cues (red and green) and the consequences of responding to those cues (protein and fat). A more realistic example of a univalent

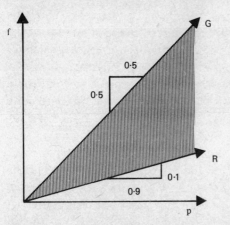

Fig. 2.9 Vectors in a consequence plane resulting from consumption of green G and red R food. The shaded area represents the realizable consequence space. (After Sibly and McFarland, 1974.)

environmental structure would be water, for which the two cues, one visual and one thermal, might relate directly to the hydrating and thermal consequences of drinking the water. In the case of the univalent food matrix (equation (2.5)), any point in the consequence space can be reached by judicious choice of red and green food. Suppose, however, that the constitution of the green food were 50% protein and 50% fat, and that of the red food were 90% protein and 10% fat. Then the environment would be *ambivalent*, viz.

$$E = \begin{bmatrix} 0·9 & 0·5 \\ 0·1 & 0·5 \end{bmatrix} \tag{2.6}$$

By choosing green food alone, the animal will move along vector G in the consequence space (Fig. 2.9), while choice of red food corresponds to vector R. Any point between these two vectors (shaded area) represents a realizable consequence, but points outside the shaded area are not realizable. Thus as a consequence of environmental ambivalence, certain regions of the consequence space are not realizable and therefore certain command states are not realizable. In terms of the command space (Fig. 2.10) there is a *cone of possible consequences* which corresponds to the shaded area (cone of realizable consequences) in Fig. 2.9. That is, for any starting point x_p, it is possible to reach any point in the command space that falls within the cone of possible consequences determined by the environmental matrix E. Only if the initial command state x_p lies within the

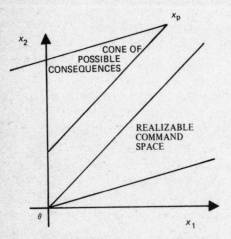

Fig. 2.10 Command space showing cone of realizable initial commands, and possible consequences of starting at an unrealizable command. (After Sibly and McFarland, 1974.)

realizable command space (Fig. 2.10) can a goal state θ be reached. Outside this space, the trajectory that results from the animal's behaviour may come near to the goal state θ but it can never reach it, because it is confined within the cone of realizable consequences, as illustrated in Fig. 2.10. The goal state θ may be any command state but it is conveniently represented as the origin of the command space. The general point is that, as a result of environmental ambivalence, certain consequences of behaviour, and therefore certain goals, cannot always be attained from a particular initial state x_p. This realizability problem has implications for the concepts of observability and optimality, which are discussed in later chapters.

2.4 Summary

The animal's *physiological state* can be represented as a point in a multidimensional *physiological space*, which has independent axes and is delimited by a *lethal boundary*.

Displacements of physiological state, induced by environmental factors, are opposed by processes ranging from slow acclimatization to fast physiological reflexes and behavioural adjustments. These processes combine vectorially, and are therefore complementary.

The *adaptation space Q* (defined in terms of limiting velocity vectors) represents the degree to which the behavioural, regulatory and acclimatory processes combine to oppose environmentally induced displacements.

The *adaptation theorem* states that the animal will die if the rate of physiological displacement extends beyond the boundary of the adaptation space.

The *regulatory space R* is the space (in terms of velocity vectors) within which the state must remain (when no behaviour is possible), if a particular state of acclimatization is to be maintained. Similarly, the *behaviour space B* is the space within which the state must remain, in the absence of physiological regulation. The *displacement space D* is a combination of R and B, and represents the extent to which rapid processes alone can maintain physiological stability.

From the fact that the processes that oppose physiological displacements are vectorially additive, it can be deduced that the acclimatized state provides the origin of the displacement space. In situations in which behaviour is the prime consideration, it is convenient to call this space the *command space* because displacements of the physiological state from the local (acclimatized state) origin constitute the actuating variables, or commands, for the mechanisms controlling behaviour.

The *ideal command space* is a vector space of finite dimensionality, in terms of which every possible command vector can be represented.

The *minimal command space* is the space of minimum dimensionality in which it is possible to represent the consequences of the animal's behaviour.

The *dimensionality theorem* states that, provided certain conditions are met, it is always possible to represent an animal's command state in a natural space of known dimensions (i.e. the minimal command space).

The consequences of behaviour are said to be *univalent* when there is a one-to-one relationship between environmental cues and the consequences of responding to those cues. If, however, the consequences of responding to a single cue related to more than one dimension of the command space, then the situation is *ambivalent*.

The *cone of possible consequences* is that subspace of the minimal command space which it is possible to reach from a given initial command. Its shape and size depend upon the degree of ambivalence in the consequences of behaviour. The *realizable command space* is the region of the command space from which a command goal can be reached, under the prevailing conditions of environmental ambivalence.

3 External factors in motivation

There are three basic ways in which changes in the external environment influence the state of an animal (see section 2.3). Changes in the environment may (1) directly influence the animal's physiological state, (2) alter the relationship between behaviour and its consequences, and (3) provide stimuli which the animal evaluates perceptually. The first two of these influences are discussed in Chapter 2; this chapter is concerned primarily with the third influence, the animal's evaluation of external cues.

3.1 The significance of environmental cues

An environmental change may have a direct physiological effect upon an animal and at the same time have significance as a cue. For example, an increase in environmental temperature induces increased water loss in many species, either by direct evaporation of water from the body surface or by stimulating evaporative cooling mechanisms, such as sweating in mammals, or pulmocutaneous evaporation in birds. These physiological responses are induced by thermoreceptors in the central nervous system (Benzinger, 1969; Richards, 1975). At the same time the environmental change may be detected by peripheral thermoreceptors, perceived as a thermal stimulus, and evaluated as a thermal cue which induces a behavioural response (Benzinger, 1969). Pigs, for example, are particularly sensitive to ambient temperature changes and will alter their behaviour accordingly (Baldwin and Ingram, 1967). Budgell (1970, 1971) showed that both rats and pigeons respond to temperature changes by drinking in anticipation of any thermally induced water loss. Thus the behavioural response will generally serve to ameliorate the physiological effects of the environmental change, as discussed in Chapter 2.

The perception of environmental stimuli and their evaluation by the animal involve many complex neuro-physiological processes, which do not concern us here. Although many of these processes are not fully understood, we feel that we can make a number of assumptions which will be sufficient for our present purposes.

Firstly, sensory capabilities are always limited, and no animal perceives every aspect of an environmental situation. This means that there will be certain environmental events which have no role as cues, even though they may directly affect the animal's state. For example, certain types of radiation may alter an animal's physiological state, but may not be detected by the animal's sensory apparatus. Other aspects of the environment may not be perceived by the animal, even though they may affect the consequence of the animal's behaviour. For example, as a consequence of feeding, animals may ingest undetected vitamins, poisons and other substances which can cause marked changes in physiological state (Rozin and Kalat, 1971; McFarland, 1973).

A second assumption is that certain specific aspects of the stimulus situation may have a uniquely powerful significance for the animal. There are many, well documented examples of such ethological sign stimuli. However, a corollary of this situation is that animals may make mistakes in interpreting environmental stimuli. A famous example is Lack's (1943) observation that male European robins (*Erithacus rubecula*) will respond to a bunch of red feathers placed within their territory, as if it were a rival male with its typical red breast. Stuffed robins lacking this sign stimulus were ignored. In addition to such innate bias, animals learn to attend to particular aspects of environmental stimuli, and to ignore others (Mackintosh, 1973).

A third assumption is that animals generally make the most of the available stimuli that they are capable of detecting. Even though we may know little about the perceptual processes involved, we assume that somehow the animal interprets the stimulus situation in a way that is concordant with its own interests and knowledge. Evidence from studies of the perceptual abilities of animals (e.g. Hailman, 1977) and from ways in which stimuli are used in such tasks as bird navigation (Keeton, 1974) tends to support this view.

Finally, we assume that the evaluation of environmental cues is, at least to some extent, quantitative, so that some stimuli may be said to be stronger, more relevant, or more effective, than others. Here we enter into an area of inquiry that is directly relevant to our theoretical framework.

3.2 The cue space

In Chapter 1 we follow McFarland and Sibly (1975) in representing the motivational state of an animal (the causal factor state) in a causal factor space with axes corresponding to the internal and external causal factors for behaviour, as illustrated in Fig. 1.4. In Chapter 2 we discuss the organization of the internal causal factors, and show that these may, under certain conditions, be treated separately in a command space. McFarland and Sibly (1975) proposed that the external causal factors be treated in an analogous manner. They used the term *cue strength* to denote the effectiveness of an external stimulus, and suggested that a particular stimulus situation will have a number of dimensions, each generating a different cue strength. Since we wish to make no assumptions about the way these cues interact, we follow McFarland and Sibly (1975) in combining the various cue strengths in a *cue space*. This space, which is somewhat analogous to the command space, has an independent axis for each cue. A point in the cue space represents the *cue state* associated with a particular environment, at a particular time.

As an example, let us consider the investigation of egg retrieval by herring gulls, carried out by Baerends and his coworkers (Baerends and Kruijt, 1973). They studied the effectiveness of various dummy eggs in eliciting the response of retrieving an egg placed on the nest rim. They use a 'titration' method for quantifying egg preference which is illustrated in Fig. 3.1. Prior to the experiment, the dummy eggs were graded according to size, because it was known that, when given a choice between a large and a small egg, the birds will generally retrieve the larger egg. The size of an egg was classified according to the area of its maximum projection onto a plane surface and arranged in a numbered size series R. The first part of each test, as illustrated in Fig. 3.1, is designed to determine the position preference of each bird. In I(a) two equal-sized eggs of normal shape and colouration are placed on the nest rim. The right-hand egg is chosen by the bird; in the next trial I(b) this is replaced by a smaller egg. Again the right-hand egg is chosen. This time I(c) the left-hand egg is replaced by a larger one. Now the bird switches its choice to the left-hand egg. The change in position preference occurs when the ratio r between the maximal projection areas is between 1·3 (I(b)) and 1·5 (I(c)). This conclusion holds for subsequent tests with dummy eggs of normal shape and colour. Once the position preference of a

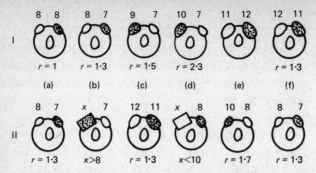

Fig. 3.1 The 'titration' method for determining the value of an egg dummy. The circle represents the nest with one egg in the nest bowl and two dummies on the rim. The code numbers 7, 8, 9, 10, 11, 12 refer to the dummies of the size series R shown in Fig. 3.2; x is the model to be measured; r is a way of indexing the R series and is defined as the ratio between maximal projection surfaces of the dummies on the nest rim. The black dummy is always the preferred one.

I. Determination of the value of the position preference. I(a) shows that the right site is preferred. This preference remains when dummy 8 is replaced by the smaller dummy 7 (I(b)) but can then be overcome by replacing 8 by 9 (I(c)); this sequence shows that the value of the position preference lies between $r = 1\cdot3$ and $r = 1\cdot5$: this conclusion holds when another pair of dummies with the same ratio is used (I(f)). Control test I(e) shows that the preferred size for this gull exceeds size 11.

II. (b) Determination of the r-value of model x. Tests II(a), II(c) and II(f) show that the position preference has remained unchanged. Tests II(b) and II(d) indicate, in combination with the preceding and succeeding tests, that the r-value of x is between those of the models 8 and 10 of the reference size series. (After Baerends and Kruijt, 1973.)

bird has been determined in terms of the r-value, then a dummy of unnatural shape or colour can be tested against this. This is exemplified in series II of Fig. 3.1. In II(a) the position preference is confirmed, and in II(b) a test egg x is introduced. The birds' responses in this series of tests indicates that the r-value of x is between those of the dummy eggs 8 and 10 of the reference size series R.

A summary of the results of many such trials is illustrated in Fig. 3.2. This shows how the value of a dummy with respect to the standard size-ranking series R is affected by various changes in features, namely (a) changing the egg shape into rounded-edge block shapes; (b) omitting the spotted pattern on standard brown dummies; (c) changing the standard brown background for a green one; (d) adding a speckled pattern to the green dummies. All changes were carried out on dummies of different sizes. These results show that the birds respond to various features of the dummy eggs, namely the size,

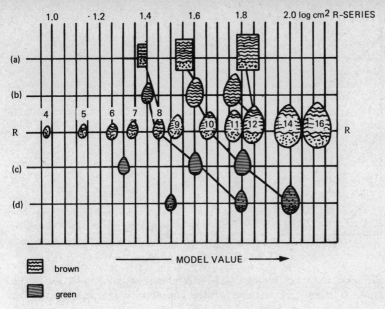

Fig. 3.2 The average values found for various dummies with respect to the reference size series R. The position of different types of dummies (brown, speckled, block-shaped; brown, unspeckled, egg-shaped; green, unspeckled, egg-shaped; green, speckled, egg-shaped) each of different sizes was determined with the method described in the legend of Fig. 3.1. The code numbers 4 to 16 in the R series stand for, respectively, 4/8 to 16/8 of the linear dimensions of the normal egg size (8 = 8/8). The maximal projection surfaces (cm^2) of the eggs of the reference series have been plotted (egg centres) along the logarithmic scale of the abscissa. Equal distances between points on this scale imply equal ratio values. (After Baerends and Kruijt, 1973.)

shape and colouration of the eggs. We might like to think of these as different cue strengths. On this basis we could represent the situation in terms of the cue space illustrated in Fig. 3.3. However, these results also show that the various features of a dummy combine in their effect upon the response. Each feature adds a specific quantitative contribution which is independent of the contribution of other features.

This finding is in agreement with the 'law of heterogeneous summation' first propounded by Seitz (1940). The law states that the independent and heterogeneous features of a stimulus situation are additive in their effects upon behaviour. There have been a number of quantitative demonstrations of the law of heterogeneous summation (e.g. Rilling, Mittelstaedt and Roeder, 1959; Heiligenberg, 1966,

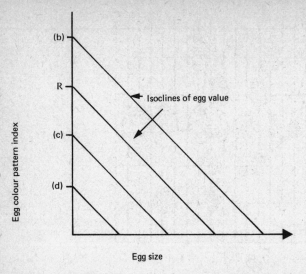

Fig. 3.3 Isoclines in cue space for the case in which features (colour and size) of the stimulus complex (the egg) are additive in their contribution to cue strength (egg value).

1969; Leong, 1969; Heiligenberg, Kramer and Schultz, 1972). However, stimulus summation usually operates within a relatively restricted range of cues, such as those pertaining to egg recognition, and many features of the stimulus situation will have entirely independent effects upon behaviour. In some instances there may be interaction between cues. Thus Curio (1975), working on the antipredator behaviour of the pied flycatcher (*Ficedula hypoleuca*), obtained some evidence of multiplicative relationships between stimulus aspects of models of two natural predators, the red-backed shrike (*Lanius collurio*) and the pigmy owl (*Glaucidium passerinum*).

Returning to the case of simple stimulus additivity, such as that illustrated in Fig. 3.3, the axes of the cue space can be combined into a single axis. Thus all those additive cues involved in egg recognition could be combined into a single cue strength for egg recognition features and represented along a single dimension. This may be set against some other independent feature of the egg-retrieval situation, such as distance from the nest, which would be represented along a different dimension. Presumably, there would then be a set of motivational isoclines joining points of equal tendency to retrieve an egg from outside the rim of the nest, as illustrated in Fig. 3.4.

In general, we should make no assumptions about the construction of the cue space, and should allow for a variety of possibilities. In

External factors in motivation   43

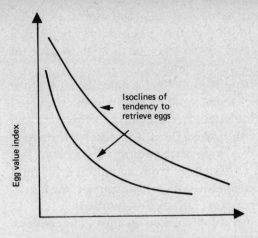

Fig. 3.4 Isoclines in causal factor space indicating the relationship between egg value and the inverse of egg distance in contributing to the strength of the tendency to retrieve the egg.

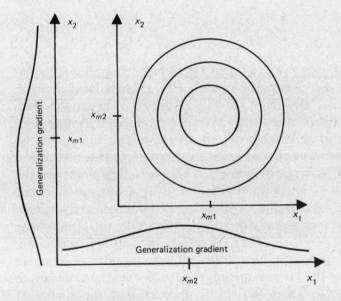

Fig. 3.5 Isoclines in cue space for a case in which there is generalization along each axis. The bell-shaped curves indicate the cue strength to each point along the axis. x_{m1} and x_{m2} are the points of greatest cue strength. The concentric circles (inset) are the resulting isoclines. Cue strength increases toward the centre of the circle. Perfect circles result from identical generalization functions, but this will not generally be the case.

some cases the cue strength may have an all-or-none aspect, as in the response of the male silk moth to the pheromone bombykol (Schneider, 1969). In other cases there may be a high degree of stimulus generalization, as illustrated in Fig. 3.5.

An interesting suggestion (Pring-Mill, 1979) is that the cue space can be recalibrated as a result of a discrepancy between the expected and obtained consequences of behaviour. According to this view, an animal would rate more highly the cues relevant to the performance of an activity for which it has a high expectation of success. The nature of the feedback it then experiences from actually performing the activity acts upon this assessment, either lowering or raising it, according to whether or not the expectations are fulfilled.

The cue space is thus subjectively calibrated in units of feedback discrepancy. Pring-Mill suggests that large recalibrations of cue space causes switches of attention, which result in changes in behaviour. Displacement activities, and other instances of behavioural disinhibition, can be accounted for by representing McFarland's (1966) frustration-attention hypothesis in a state-space framework.

3.3 Time cues

The behaviour of many species can be influenced by endogenous clocks which may be circadian (Aschoff, 1965), lunar (Hauenschild, 1960) or circannual (Berthold, 1974) in period. The endogenous clock may directly influence the animal's physiological state. For example, in migratory white-crowned sparrows the timing of the departure from the wintering grounds in the Arizona-Mexico region is fairly critical. Arrival too early at the breeding areas in the north-western U.S.A. and Canada may result in a disastrous exposure to a period of cold weather. On the other hand, late arrivers may find the best territories already occupied. Moreover, the female may not have time to rear her offspring before the autumn migration. Climatic variables generally have little influence upon the timing of migration (Klein, 1974), presumably because they are prone to change from year to year. Rather, it appears that the timing of migration in white-crowned sparrows is controlled by photoperiodic factors combined with an endogenous clock, and takes place through the regulated release of hormones (Farner and Lewis, 1971). Although temporal factors often have a direct influence upon physiological state, and may therefore alter the animal's command state, in some cases it may be better to consider temporal factors as external cues. Many animals are able to learn to adjust their behaviour on the basis of time. For example, herring gulls that forage on refuse tips seem to time their arrival to coincide with the fairly limited periods of the day that food is likely to be available (McFarland, 1977). They do not show up on Saturdays and Sundays, but they do appear on public holidays. This indicates that the birds are not using direct external cues, such as the sound of the refuse trucks, but rely instead upon an internal timing mechanism. A similar reliance can be demonstrated in relation to the tidal cycle.

Whether we choose to regard temporal factors as cues or commands makes no difference to our present analysis. As we saw in Chapter 1, the axes of the cue space are combined in a Cartesian product with those of the command space. In combining external and internal factors in this way we are making no assumptions about the mathematical functions by which the internal and external factors relevant to a particular behaviour are related. We regard this question as an entirely empirical matter, which is manifested in the shape of the motivational isoclines. For the purposes of illustration, however, we can imagine that the tendency for a gull to forage at the refuse tip depends jointly upon its state of hunger and upon a time cue which represents the animal's estimate of the likelihood of obtaining food at a particular time. This hypothetical situation is illustrated in Fig. 3.6.

For an animal to behave appropriately in interactions with its environment, information about that environment must be incorporated into the mechanisms responsible for the behaviour. Oatley (1974) has argued that, although it is possible to discuss the interactions of circadian rhythmicity and motivation purely in terms of physiological and biochemical processes, it is also possible to postulate the existence of emergent properties, in the sense that oscillations inherent in motivational feedback processes can be pressed into services as a working model of terrestrial rotation. The word 'model' is used here to denote an animal's representation of the environment. The advantage of such models is that they enable animals to anticipate environmental changes and to develop elaborate strategies of appetitive behaviour. In this view, learning is seen as the process of constructing successively more effective representations of the environment (Minsky, 1968). Just as artificial clocks are models of the rotation of the earth and the consequent alternation of light and darkness, so too are the oscillators underlying circadian rhythms

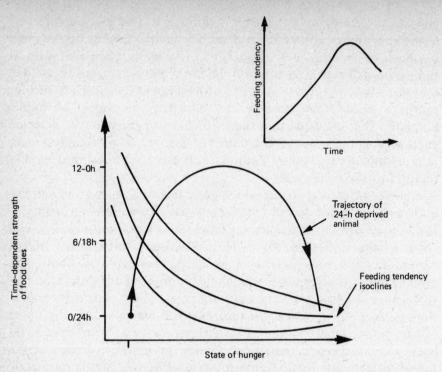

Fig. 3.6 Motivational isoclines in a causal factor space in which one aspect of the animal's state is time dependent, as explained in the text. In this example, the hunger trajectory follows a ∩ function over a 24-h period. This produces changes in feeding tendency with time, as is shown in the inset graph.

(Oatley, 1974). Such a model can serve not only to synchronize various physiological processes, but to entrain the behaviour of the animal to the rhythmic characteristics of its niche. For example, there is growing evidence that sleep is not well described in terms of a homeostatic mechanism involving recuperation from periods of wakefulness (Meddis, 1975). Sleep is better understood in terms of ecological variables (Zepelin and Rechtschaffen, 1974; Allison and Cicchetti, 1976). Specifically, sleep patterns seem to correlate with niche characteristics which have circadian components, such as foraging opportunities, climatic factors and danger levels. Products of fatigue may have a modulating effect upon the basic sleep cycle, but the circadian cycle probably serves to distribute periods of recuperation to the times of day which would in any case be unsuitable for activity (Oatley, 1974).

Another more sophisticated feature of circadian rhythms as models of terrestrial rotation is their role in animal navigation. There is considerable evidence that some type of circadian clock is used by animals of many species to provide cues to the relationship between the positions of heavenly bodies as seen from different positions on the earth's surface (Matthews, 1965; Emlen, 1974; Keeton, 1974).

Evidence favouring the view that oscillatory physiological processes provide the basis for a model capable of providing time cues comes from studies showing that animals can maintain out-of-phase rhythms simultaneously. Oatley (1971) showed that although feeding and drinking in rats are normally highly synchronized, with a marked circadian rhythm dictated by the feeding pattern, the feeding and drinking rhythms can be experimentally disassociated. Rats injected with varying quantities of sodium chloride at different times of day drink exactly the same amount of water in response to a given osmotic stimulus, independently of the phase of the circadian cycle. Moreover, rats fed only once each hour maintained a normal circadian pattern of water intake. Thus although the three activities of eating, drinking and sleeping all have systematic circadian rhythms and need to be synchronized together, this does not appear to be done by one activity simply stimulating, permitting or inhibiting the others. Rather it appears that there are independent oscillators which synchronize with one another by being sensitive to interactions between their respective systems (Oatley, 1974). However, Edmonds and Alder (1977) showed that rats experimentally fed according to a predetermined schedule could simultaneously maintain a 24-h rhythm and a 25-h rhythm of feeding. This result shows not only that there are different oscillators which are simultaneously active but that the oscillatory mechanisms are capable of providing time cues that can be utilized by the animal, even to control behaviour within a single system.

We are not advocating, at this stage, any ambitious foray into cognitive aspects of stimulus evaluation or of models of the environment. However, we think that it is important to recognize that such systems probably do exist. The ultimate viability of a theory or representation of motivational mechanisms depends upon the adaptability of the theory with respect to new evidence and new concepts. We have tried to formulate our representation of the animal's perceptual world to take account of the high probability that future research will show that some animals are capable of developing cognitive models of their environment which can provide cues for behaviour. It does not matter from the point of view of our theory whether the cues emanate directly from environmental stimuli, as in the case of simple sign stimuli, or whether they come from an animal's internal model, as in the case of time cues. What matters is how their cues interact with command variables, which is an empirical question. How the experimenter is to measure and evaluate these interactions is the subject of the next chapter.

3.4 Summary

We follow McFarland and Sibly (1975) in treating external causal factors in a manner analogous to that of internal factors. Thus we use the term *cue strength* to denote the contribution of an external stimulus to the animal's causal factor state. Making no assumptions about the way in which different aspects of the external stimulus situation combine and interact, we represent the independent features of the external environment along different axes of a multidimensional *cue space*. A point in the cue space represents the *cue state* associated with a particular environment, at a particular time.

We envisage that changes in cue state can occur as a result of time cues. Thus we postulate that in addition to their influence upon the animal's internal state, endogenous clocks can provide cues which the animal can utilize in assessing the significance of external stimuli. Since the command and cue states combine to give the causal factor state, it makes little mathematical difference whether temporal factors are viewed as internal or external cues. However, in so far as endogenous clocks provide a model employed by the animal to represent certain aspects of the external world, we prefer to represent time cues as contributions to the animal's evaluation of the external environment. This attitude opens the way for the incorporation of cognitive models into our schema, and we feel that this may be important in the future, when we may hope to have a greater understanding of cognitive processes in animals.

4 The interaction of internal and external factors

4.1 Introduction

In this chapter we consider the question of how cues and commands combine to produce a behaviour tendency. A major cause of confusion in this area has been the assumption that the measurement scale that the experimenter uses for these variables corresponds to the scale used by the animal (Houston and McFarland, 1976). In arguing against this view we describe a general approach to measurement which we believe to be of wide applicability in the behavioural sciences.

A classic example of the combination of internal and external factors is provided by the work of Baerends *et al.* (1955), already mentioned in Chapter 1 (see Fig. 1.9). The size of the female, and the male's sexual motivation, jointly determine the male's courtship behaviour. The shape of the isoclines in the two-dimensional causal factor space appears to be hyperbolic, suggesting that cues and commands are multiplied to produce the behavioural tendency. Houston and McFarland (1976) questioned the validity of this interpretation and introduced the term 'combination rule' to describe the relationship between causal factors and behavioural tendency. We will argue that the combination rule cannot be established without considering the measurement of the causal factors.

4.2 The problem of measurement

Measurement consists of assigning numbers to objects in such a way that the relations between the numbers reflect the relations between the objects. A formal presentation of this idea can be found in Pfanzagl (1968) or Krantz, Luce, Suppes and Tversky (1971). For our purposes it is sufficient to note that there may be many numerical assignments which are consistent with the empirical relations between the objects of investigation. Indeed, the 'flexibility' of the assignment can be used to define the measurement scale, as is shown in Appendix 1. This ideal can be illustrated by considering the musical recordings that make up the 'Top Twenty'. The numbers one to twenty that are given to these records merely indicate the order of the record, in terms of numbers sold. Any set of twenty numbers could be used as long as they were assigned to the records in such a way that the record that was bought the most received the smallest number, and so on. This amounts to saying that the numbers one to twenty are only unique up to a transformation that preserves their order. The set of such transformations roughly defines what are known as ordinal scales, in which only the ranking of the objects is represented.

Although this example may seem trivial, it is important in that it makes clear the distinction between the objects themselves and the numbers that we assign to them. Furthermore, it illustrates the fact that many numerical assignments convey the same information. Both these points are central to our treatment of the problem of cues and commands, to which we now return.

The earlier chapters have described the idea of causal factor space, which combines all the axes of cue space and command space. The question of how internal and external factors interact can thus be answered by specifying how cues and commands combine to produce behavioural tendency. This is not as easy to do as it sounds, because of the problems of measuring these variables. A difficulty arises because we cannot assume that our measurements of the internal and external factors (y_i and z_j) will coincide with the calibrations by which the animal obtains the cues and commands. This point is illustrated by Fig. 4.1. It is obvious that external stimuli must be calibrated (i.e. measured) by the animal to produce cues, but the same is also true for the internal displacements. McFarland (1971) points out that such variables, being analogous to mechanical displacements, cannot be

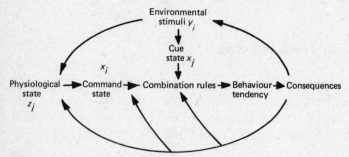

Fig. 4.1 Outline of relationships between the internal and external factors governing behaviour.

Table 4.1 Analogous variables and parameters (from McFarland, 1971).

	Mechanical	Electrical	General
Rate variables	force f velocity u	voltage v current i	effort e flow f
State variables	displacement x momentum p	charge q flux λ	displacement h momentum p
Parameters	mass M compliance K friction B	inductance L capacitance C resistance R	inductance L capacitance C resistance R
Power	$HP = fu$	$P = vi$	$P = cf$
Potential energy	$E_p = \frac{1}{2}Kf^2$	$E_p = \frac{1}{2}Cv^2$	$E_p = \frac{1}{2}Ce^2$
Kinetic energy	$E_k = \frac{1}{2}Mu^2$	$E_k = \frac{1}{2}Li^2$	$E_k = \frac{1}{2}Lf^2$

'actuating' variables. They must be measured by the animal to produce 'effort' variables, which we call commands in this context (see section 2.2). These commands are analogous to voltages in electrical systems, as is shown in Table 4.1. The link between a voltage and an electrical displacement is a capacitance, and we follow McFarland (1970) in borrowing this terminology for the calibrating parameter of a physiological displacement.

Using the framework of Fig. 4.1, we will now discuss the results of Baerends *et al.* in more detail. To facilitate the discussion we introduce the following notation:

Let a given external stimulus be denoted by y_i
$$i = 1, \ldots, n$$

Let a given internal displacement be denoted by z_j
$$j = n+1, \ldots, n+m$$

The observer's measurement of y_i will be denoted by $m_i y_i$ and his measurement of z_j will be denoted by $m_j z_j$.

At the risk of oversimplification, we assume for the moment that each y and z corresponds to a causal factor, i.e.

$$x_i = b_i y_i \qquad b_i \text{ in B}$$

and

$$x_j = c_j y_j \qquad c_j \text{ in C}$$

where x_i and x_j are causal factors (cues and commands respectively), B is the set of calibrations of stimuli and C is the set of calibrations of commands (i.e. the set of capacitances).

In the case of the courtship of the male guppy, we want to know how $x_1 = b_1 y_1$ and $x_2 = c_2 z_2$ combine to produce a behavioural tendency. A result such as that of Baerends *et al.*, however, only tells us how $m_1 y_1$ and $m_2 z_2$ combine. This will be sufficient if the animal uses the same calibration as we do, but will be misleading if m_1 or m_2 differs from b_1 or c_2. For example, if b_1 and c_2 are such that x_1 is the log of $m_1 y_1$ and x_2 is the log of $m_2 z_2$, then x_1 and x_2 would combine additively to determine the behavioural tendency. As long as our measurements of causal factors are only on ordinal scales, such alternative calibrations will be possible, and it cannot be said whether causal factors combine by multiplication or addition. This claim is established in Appendix 1, together with a proof that if measurement on interval scales is achieved then the additive and multiplicative combination rules can be distinguished.

4.3 The conjoint measurement approach

The argument so far can be summarized as follows: it cannot be assumed that the shape of the isoclines in causal factor space tells us the nature of the combination rule, because this shape depends on the scaling of the causal factors. The shape is only significant if the scale is an interval one, i.e. one that preserves relative distance. Baerends *et al.* admit that their calibration procedure, which constructs a measure of the male's sexual state on the basis of his colour patterns, results in no more than an ordinal scale: 'However, these and other criticisms (of the calibration procedure) are chiefly concerned with the exact distances between successive marking patterns on the scale, not primarily with the order in which the patterns develop with increase of internal stimulation. This order, which is the principal basis for our further conclusions, seems fairly well established' (p. 306).

Given this limitation, the isoclines must be thought of as establishing no more than an ordering of the causal factors, on the basis of their 'competitiveness' in gaining access to the behavioural final common path.

The conjoint measurement approach to the difficulties created by the interdependence of the calibrations and the combination rule involves tackling both these issues simultaneously. The idea is presented in detail by Krantz and Tversky (1971), who use the term 'composition rule' to describe the relationship between the dependent and independent variables; our combination rules are examples of what they call composition rules, as are the well known gas laws of Hooke and Boyle. Rather than asking whether or not a given composition rule holds, an attempt is made to establish properties of the ordering of the independent variables that will distinguish between various possible composition rules. As Krantz and Tversky put it: 'The problem... therefore is not whether a specified functional relation holds among several (independently measured) variables, but rather whether there exist scales of measurement (for both the dependent and the independent variables) that satisfy the proposed composition rule' (p. 152).

Krantz and Tversky consider the four simple ways in which three variables can combine:

$A + P + U$	additive rule
$(A + P)U$	distributive rule
$AP + U$	dual-distributive rule
APU	multiplicative rule

We have already shown (p. 52) that, for positive scale values, the additive and multiplicative rules cannot be distinguished. Krantz and Tversky (1971) point out that the other polynomials can be separated on their ordinal properties. For example the additive rule is characterized by the fact that all the variables satisfy joint concurrence (defined below). Two conditions, known as concurrence and double cancellation, are necessary for any of the rules to hold. We define these conditions below, using the following notations:

A, P, U are motivational states, with a, b, c, in A; p, q, r, in P and u, v, w, in U respectively. A, P, U are scales, i.e. numerical functions defined on **A**, **P**, **U** respectively. $r(a, p, u)$ denotes the tendency resulting from state (a, p, u).

4.3.1 Concurrence*

A variable is said to be concurrent if the ordering of the responses that arise from changing the value of the variable is the same whatever the values of the other variables may be (the values of the other variables are held constant during each determination of an ordering).

Formally, **A** is concurrent with respect to **P** and **U** whenever

$r(a, p, u) \geq r(b, p, u)$ if and only if

$r(a, q, v) \geq r(b, q, v)$ for all a, b in **A**, p, q in **P** and u, v in **U**

In other words, if a results in a stronger response than b for one given set of states, it will result in a stronger response for all other sets of states.

Similarly, two variables are jointly concurrent if the ordering they jointly impose on the responses does not depend on the value of the

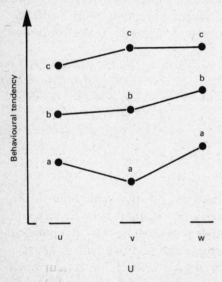

Fig. 4.2 Plot of **U** against the tendency parameterized by **A**. (After Houston and McFarland, 1976.)

* Krantz and Tversky (1971) call this condition 'Independence' but this term might be confused with 'independent variable' (e.g. 'The essence of independence is that the ordering of the dependent variables can be used to order some of the independent variables... in a manner that does not depend on the remaining variables' (Krantz and Tversky, 1971, p. 155).

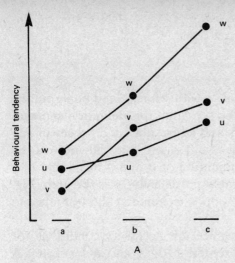

Fig. 4.3 Plot of **A** against the tendency parameterized by **U**. (After Houston and McFarland, 1976.)

third variable, i.e.

A and **P** are jointly concurrent with respect to **U** whenever
$r(a, p, u) \geq r(b, q, u)$ if and only if
$r(a, p, v) \geq r(b, q, v)$

The concurrence of a variable can be verified graphically, as is shown in Figs. 4.2 and 4.3. The causal factor **A** is concurrent with respect to **U** if a plot of **U** against tendency, parameterized by **A**, results in uncrossed lines when points corresponding to a given member of **A** are connected. This may sound complicated, but Fig. 4.2 should make the principle clear. In this case, **A** is concurrent with respect to **U**, but as can be seen from Fig. 4.3, **U** is not concurrent with respect to **A**. We will say that **A** and **P** are concurrent if each is concurrent with respect to the other.

4.3.2 Double cancellation

This condition is not as easy to describe as concurrence. It has some resemblance to transitivity in that it enables a relation to be deduced from a pair of relations. The rough idea is that if we have some comparisons of response strength for states drawn from **A** and **P**, we might expect a combination of a state from **A** that was always on the 'winning' side with a similar state from **P** would 'beat' a combination based on states that were always on the losing side. This can be put rigorously as follows: **A** and **P** satisfy double cancellation if

$r(a, q) \geq r(b, r)$ and $r(b, p) \geq r(c, q)$

implies

$r(a, p) \geq r(c, r)$

If we identify being on the left of $\geq$ with winning and being on the right of $\geq$ with losing, then it can be seen that this definition captures the rough idea sketched above. a and p are always on the winning side, q and b appear on both sides, and c and r are always on the losing side. Consequently the combination of a and p beats that of c and r. The same idea can be illustrated graphically, as is shown in Fig. 4.4. In this figure, 'winning' corresponds to being at the top left end of the arrows.

Double cancellation and concurrence are necessary conditions for the existence of an additive representation of **A** and **P** by scales A and P respectively. A representation is additive if the ordering of the response strengths produced by **A** and **P** corresponds to the ordering created by the sum of the scale values of the states. To be precise, **A** and **P** admit an additive representation, if they can be given values defined by real-valued functions f and g respectively such that

(i) $f(a) + g(p) \geq f(b) + g(q)$ if and only if

$r(a, p) \geq r(b, q)$

(ii) $f(a) \geq f(b)$ if and only if

$r(a) \geq r(b)$

(iii) $g(p) \geq g(q)$ if and only if

$r(p) \geq r(q)$

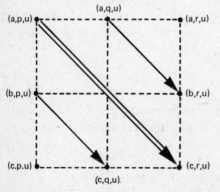

Fig. 4.4 Graphical illustration of the double cancellation axiom. The thin arrows represent the $\geq$s of the premise, the thick arrow represents the $\geq$ of the implication. (After Houston and McFarland, 1976.)

The other requirements for this existence result are given in Appendix 2.

The existence of an additive representation implies double cancellation, as the following argument shows.

If there is an additive representation of **A** and **P** by functions f and g respectively, then

$$r(a, q) \geq r(b, r) \text{ if and only if } f(a) + g(q) \geq f(b) + g(r)$$

and

$$r(b, p) \geq r(c, q) \text{ if and only if } f(b) + g(p) \geq f(c) + g(q)$$

Adding the terms on the right of 'if and only if' yields

$$f(a) + g(q) + f(b) + g(p) \geq f(b) + g(r) + f(c) + g(q)$$

$g(q) + f(b)$ can be subtracted from each side to give

$$f(a) + g(p) \geq f(c) + g(r)$$

which by the existence of an additive representation, is equivalent to

$$r(a, p) \geq r(c, r)$$

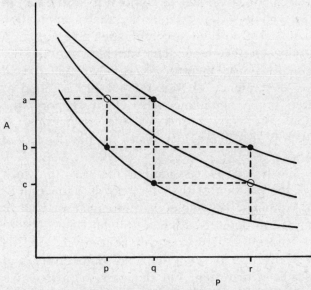

Fig. 4.5 The isoclines of Baerends *et al.* (see Fig. 1.9) considered with respect to double cancellation. The two outside isoclines provide starting points (black dots) for the double cancellation axiom. The dotted lines are constructed at right angles to the axes. The points at which the outer construction lines cross (open circles) should lie on an isocline if double cancellation holds, which they do in this case. (After Houston and McFarland, 1976.)

The implication of this section for the results of Baerends *et al.* is that if the response strengths are to be represented as an additive function of female size and male sexual state, then the double cancellation condition must be satisfied. This is also true if a multiplicative representation is required, because the additive and multiplicative representations are equivalent (see section 4.2). Replacing the inequalities in the double cancellation condition by equalities gives a set of requirements for the isoclines (see Appendix 4.3 for details).* One such condition is tested graphically in Fig. 4.5. If all these double cancellation conditions are always met by the isoclines, then we can infer that the courtship behaviour of the male guppy can be represented as an additive (or a multiplicative) combination of the female's size and the male sexual state.

4.4 The functional measurement approach

Anderson (1978) describes a method of deciding between additive and multiplicative representations on the assumption that the responses provide a linear scale of tendency. His approach, which he calls 'functional measurement', is similar to conjoint measurement in that it tackles both the scaling of the variables and their combination rules. The basic idea is that a postulated combination rule implies a certain relationship between the dependent variables and the response, given that the response measure is a linear function of response strength (i.e. of tendency). If this relationship is found to hold it achieves three things simultaneously: (1) it supports the proposed combination rule; (2) it justifies the assumption of response-measure linearity; (3) it provides scales for the dependent variables.

This idea can be illustrated by considering the way in which Barbary doves change the position of their feathers as a function of two state variables. Although the whole thermoregulatory control system has been analysed in detail by McFarland and Baher (1968) and McFarland and Budgell (1970), we will concentrate on the feather response as a function of hypothalamic temperature and degree of food or water deprivation. On the basis of a negative feedback system, McFarland and Budgell suggested that the following relationship would hold:

$$F = T_h \times K_h$$

* Luce and Tukey (1964) discuss this condition for indifference curves.

where F is a measure of feather position
   T_h is the change in hypothalamic temperature
and K_h is a parameter that depends on the state of food and water balance.

This proposed combination rule could be checked by plotting F against K_h for various values of T_h. If the combination rule is multiplicative, a 'linear fan' (Anderson, 1978) would be expected, as is shown in Fig. 4.6. Constructing this figure requires that the scale values of K_h are known, so that the various degrees of deprivation can be placed correctly along the abscissa. It might seem that no progress can be made without these values, but if the model is correct then the response to a given degree of deprivation, averaged over the temperature conditions, is a linear function of the appropriate scale value. Figure 4.6 shows that the data of McFarland and Budgell yield a linear fan. This provides a justification for the response measure F, which is based on the arbitrary feather index invented by McFarland and Baher (1968).

This argument is summarized by the linear fan theorem (Anderson, 1978), which states that if (1) a multiplicative combination rule holds,

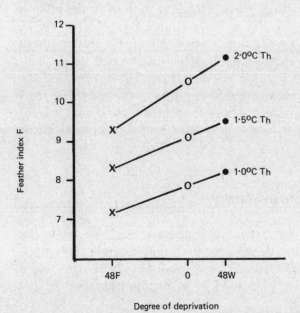

Fig. 4.6 Feather index obtained by McFarland and Budgell (1970) as a function of hypothalamic temperature (T_h) after zero (0); 48 h food (48F), or 48 h water (48W) deprivation.

and (2) the response measure is a linear scale then: (a) the appropriate plot of the response data will produce a set of diverging straight lines, and (b) the average response to a given motivational state or stimulus condition will be a linear function of the scale value of the state or stimulus.

Applications of the functional measurement approach to a number of problems are described in Anderson (1962, 1970, 1978) and Klitzner and Anderson (1977). The general approach is similar to the idea of canonical representations as defined by Pfanzagl (1968). He suggested that, given an empirical relationship between a set of variables and a response, it was reasonable to look for scalings of the variables which result in simple composition rules. Scalings which are unique up to linear transformations establish canonical representations. The idea was applied to the matching law (see below) by Houston (1977) to reach conclusions similar to those of Anderson (1978).

Anderson (1978) points out that functional measurement has the advantage over conjoint measurement of having an error theory (analysis of variance can be used in the relevant data). On the other hand, conjoint measurement requires only an ordinal response scale. In the case of guppy courtship, each behaviour pattern of the male is stereotyped, that is to say it has a 'typical intensity' (Morris, 1957). Thus each pattern provides no more than ordinal information about the male state. Even in those cases in which behaviour varies continuously, it is unlikely to provide a linear scale. This means that the application of functional measurement requires a non-linear transformation of the response measure. There is no firm theoretical basis for such transformations, so it is not clear how useful functional measurement is in such cases.

4.5 The problem of dimensionality

In our development of the problem of measurement, we made the assumption that for every dimension that the observer measures, there corresponds a causal factor. This will not always be true, and a failure to realize this will often lead to a misrepresentation of the animal's combination rules.

For example, Heiligenberg (1976) considers the contribution that the angle of eye-bar on a model of the cichlid fish *Haplochromis burtoni* makes to the behaviour of a fish that is presented with the model. Heiligenberg assumes that the absence of eye-bar corresponds to zero on the eye-bar angle dimension, but it could well be that the fish has more than one dimension on which to represent the eye-bar markings.

In some circumstances the representation of stimuli can be simplified. Suppose pigeons are presented with two out of three sorts of grain, each sort being available on an independent variable interval schedule. (A fixed interval schedule, F.I.–T sec, is one in which a response will be rewarded only after T sec have elapsed since the last reinforcement. In contrast, on a variable interval schedule, V.I.–T sec, the inter reinforcement interval varies according to some schedule with average interval of T sec.) Under these conditions, known as concurrent V.I.–V.I., it is well known that pigeons show matching behaviour (see Herrnstein (1970) for a review); that is to say

$$\frac{P_1}{P_2} = \frac{R_1}{R_2}$$

where P_i is the number of pecks delivered to schedule i and R_i is the number of reinforcements received from schedule i. Baum (1974) argues that any apparent deviation from the matching relationship means that the reinforcement values of the schedules have not been correctly assessed. In other words, matching serves to define a transformation which will establish a scale of reinforcement values. Thus the most general form of the matching law would be

$$\frac{P_1}{P_2} = \frac{S_1}{S_2}$$

where S_i represents the value of reinforcement obtained from schedule i.

Miller (1976) presented pigeons with two out of three sorts of grain on a V.I.–V.I. schedule and was able to use the results of presenting two of the possible pairs, together with the matching equation, to predict the result of presenting the third pair. As he points out, this amounts to the pigeon having a one-dimensional valuation of the three grains. The scale of grain value obtained by Miller is illustrated in Fig. 4.7.

The problem of establishing the relationship between physical dimensions and an organism's 'psychological' space in which the stimuli are represented is central to psychophysics. It therefore seems likely that developments in psychophysical theory, such as those

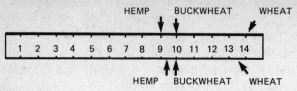

Fig. 4.7 Two scales of grain quality, one based on the amount of time that pigeons spent pecking for the various grains, the other based on the number of pecks. The quality of buckwheat was chosen as a standard and was given the arbitrary value of 10 units. (After Miller, 1976.)

described by Cunningham and Shepard (1974), will be of use in the construction of cue space.

Although this section has been largely concerned with the dimensions of cue space, considering a measure of response strength as an ordinal scale has implications for how command variables are represented. Hinde (1970) drew attention to Choy's work on the drinking behaviour of rats, shown in Fig. 4.8. 'Thirst' was assessed by the amount of water drunk in 15 min, the number of bar presses made for water, or the concentration of quinine solution that the rats would tolerate. These three ways gave different results. Hinde claims that 'In view of this lack of agreement among measures, we must conclude that the postulation of a single intervening variable is too simple a

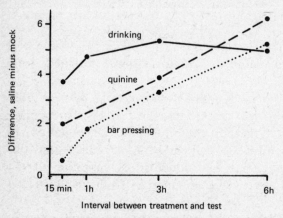

Fig. 4.8 The methods of assessing thirst in rats given 5 ml of 2 M saline by stomach tube. The ordinate indicates the amount by which the experimental test exceeds the control test, the nature of the units differing between the procedures used to measure drinking tendency. (After Hinde, 1970.)

hypothesis to account for all the changes in behaviour observed' (p. 197).

The conclusion is unwarranted if we assume that the measures do no more than order the response strengths. In fact, all the measures agree remarkably well, apart from what is presumably a ceiling effect on drinking at the 6-h interval, as shown in Fig. 4.8.

4.6 Summary

As discussed in Chapter 1, the shape of the isoclines reflects the weight the animal gives to the various causal factors. The point of this chapter is that any arbitrary measurement of the isoclines by the experimenter will not necessarily reveal, or correspond to, the animal's weightings. The importance of the approach we have outlined in this chapter is that it makes very few *a priori* assumptions about the calibrations the animal uses. We believe that these calibrations can be understood as an optimal decision-making process, designed by natural selection (see Chapter 9), although their evolutionary significance may be hidden if we impose our measurement scale on the animal.

Appendix 4.1

This appendix contains the proofs of Theorems 1 and 2.

Definitions

Measurement involves finding a scale which assigns numbers to the objects of investigation. Let the set of objects be Q and the real numbers be $\boldsymbol{R}$, then a scale is a map $m: Q \rightarrow \boldsymbol{R}$.

Ordinal scale

A scale is an ordinal scale if and only if it is unique up to monotone increasing and continuous maps of $m(Q)$ into $\boldsymbol{R}$.

Monotone increasing map

A map of an ordered set A into an ordered set B is monotone increasing if and only if $a' < a''$ implies $f(a') < f(a'')$ for all a', a'' in A.

Continuous map

A map $h: \mathbf{R} \to \mathbf{R}$ is continuous at the point a in $\mathbf{R}$ if given $\varepsilon > 0$, there is a $\delta > 0$ such that
$$|f(x) - f(a)| < \varepsilon$$
whenever
$$|x - a| < \delta$$
The map h is continuous if it is continuous at each point of $\mathbf{R}$.

Interval and ratio scales

A scale is an interval or ratio scale if it is unique up to the positive linear transformations or dilations respectively.

Positive linear transformations

The group G_p of positive linear transformations of $\mathbf{R}$ onto $\mathbf{R}$ consists of all transformations $g_{a,b}: x \to ax + b$ with a in $\mathbf{R}^+$, b in $\mathbf{R}$

$G_d = \{g_{a,o}: a \text{ in } \mathbf{R}^+\}$ is the group of dilations.

Theorem 1

If the variables and the response they produce are measured on ordinal scales and take only positive values, then it is not possible to decide between an additive and a multiplicative combination rule.

Proof

The proof is based on demonstrating that if it is possible to find a scale which allows one of the combination rules to describe the data, then the scale values can be transformed so that the other rule could be used.

Without loss of generality, assume that the measured response, r, can be represented as

$$r = v_1 + v_2 + \ldots + v_n \tag{1}$$

where the v_i are the scale values of the variables x_i. A transformation T must be found such that

$$T(r) = T(v_1) \cdot T(v_2) \ldots T(v_n) \tag{2}$$

and with the requirement that T be monotone increasing and continuous (this constraint follows from the definition of ordinal scales).

It is obvious that the exponential transformation

$$T(r) = e^r, \quad T(v) = e^v$$

is a suitable transformation.

Theorem 2

If measurement is on interval scales, it is not possible to find a transformation satisfying equations (1) and (2).

Proof

Equations (1) and (2) can be regarded as establishing a functional equation of the form

$$f(x_1 + x_2) = g(y_1) \cdot h(y_2) \tag{3}$$

It is known (Aczel, 1966) that the only continuous solutions of (3) are of the form

$$\begin{aligned} f(w) &= abe^{cw} \\ g(w) &= ae^{cw} \\ h(w) &= be^{cw} \end{aligned} \tag{4}$$

where a, b, c are constants. But as attention is now restricted to interval scales, the only permissible transformations (by definition) are positive linear ones. Thus transformations of the form of equation (4) may not be used.

Appendix 4.2

The Luce–Tukey axioms

Axiom 1 (Well-ordering)

(a) Reflexivity: $r(a, p) \geq r(a, p)$ for all a in **A**, p in **P**.
(b) Transitivity: $r(a, p) \geq r(b, q)$ and
$r(b, q) \geq r(c, r)$ imply
$r(a, p) \geq r(c, r)$

(c) Connectedness: either $r(a, p) \geq r(b, q)$ or
$r(b, q) \geq r(a, p)$ or both, i.e.
$r(a, p) = r(b, q)$.

Axiom 2 (Solutions)

For each a in $\mathbf{A}$ and p, q in $\mathbf{P}$, the equation

$$r(f, p) = r(a, q)$$

has a solution f in $\mathbf{A}$, and for each a, b in $\mathbf{A}$ and p in $\mathbf{P}$, the equation

$$r(a, x) = r(b, p)$$

has a solution x in $\mathbf{P}$.

Axiom 3 (Double cancellation)

$$r(a, q) \geq r(b, r)$$

and

$$r(b, p) \geq r(c, q)$$

imply

$$r(a, p) \geq r(c, r)$$

The fourth axiom requires some definitions: a doubly infinite series of pairs (a_i, p_i), $i = 0, \pm 1, \pm 2, \ldots$, with a_i in $\mathbf{A}$ and p_i in $\mathbf{P}$ is a dual standard sequence (d.s.s.) provided that

$$r(a_m, p_n) = r(a_q, p_r)$$

whenever $m + n = q + r$ for positive, zero or negative integers m, n, q, and r.

A d.s.s. is trivial if for all i either $a_i = a_0$ or $p_i = p_0$, in which case both hold by transfer.

Axiom 4 (Archimedean axiom)

If (a_i, p_i) is a non-trivial d.s.s., b is in $\mathbf{A}$, and q is in $\mathbf{P}$, then there exist (positive or negative) integers n and m such that

$$r(a_n, p_n) \geq r(b, q) \geq r(a_m, p_m).$$

Appendix 4.3

In this appendix an account is given of the way in which isoclines can be used to check double cancellation. Figure 4.5 shows the isoclines of Baerends et al. (Fig. 1) relabelled to make them comparable with the double cancellation axiom (p. 55). The outside isoclines provide the premises of the axiom in the following way:

As $r(a, p)$ and $r(b, r)$ lie on the same isocline,

$$r(a, q) = r(b, r) \tag{1}$$
$$\therefore \quad r(a, q) \geq r(b, r) \tag{2}$$

Similarly

$$r(b, p) = r(c, q) \tag{3}$$
$$\therefore \quad r(b, p) \geq r(c, q) \tag{4}$$
$$\therefore \quad r(a, p) \geq r(c, r) \text{ if double cancellation holds} \tag{5}$$

Equation (1) also gives us

$$r(b, r) \geq r(a, q) \tag{6}$$

and equation (3) also gives

$$r(c, q) \geq r(b, p) \tag{7}$$

so double cancellation implies

$$r(c, r) \geq r(a, p) \tag{8}$$

So, from equations (5) and (8) it follows that

$$r(a, p) = r(c, r)$$

i.e. $r(a, p)$ and $r(c, r)$ should lie on the same isocline if double cancellation holds.

5 Causal models of behaviour sequences

There are a number of ways in which empirical arguments can be brought to bear upon motivational problems. One approach is to try to identify the relevant state variables and measure their values, either physiologically or in terms of behavioural indices. Another approach is to work from sequences of behaviour, observed in either natural or laboratory conditions, and to try to account for these data in terms of postulated underlying motivational changes. In this chapter we concentrate on those methods that are most relevant to the state-space analysis of behaviour.

The most common form of attack on motivational problems is the physiological approach, which seeks to identify those systematic physiological variables responsible for activating the brain mechanisms that control behaviour, and to understand the functioning of these mechanisms in neurological terms. The task of identifying the relevant systemic variables has not proved easy. Despite an enormous amount of research, the question of what precise systemic factors are responsible for sexual motivation (e.g. Hutchison, 1978) or feeding motivation (e.g. Booth, 1978) remains controversial. In the latter case, for example, gastric contractions, gastric distension, body temperature, hunger hormones, alimentary metabolites, circulatory metabolites and oropharyngeal stimuli have all been put forward as candidates for the key variables controlling feeding. Moreover, a number of primarily psychological phenomena, such as learned aversions (e.g. Revusky and Garcia, 1970; Rozin and Kalat, 1971; McFarland, 1973) conditioned satiety (e.g. Booth, 1972, 1977) and anticipatory feeding (McFarland, 1970; Mogenson and Calaresu, 1978) are known to be implicated in feeding control, so it may be that systemic physiological factors are less important than is generally assumed.

From the point of view of quantitative ethology it is not clear that the exact physiological details of a system are very important. It may be more useful to understand the principles underlying the control of behaviour rather than the details of the implicated hardware.

Although behavioural models should not imply physiological nonsense, it is not necessarily advantageous to rely upon physiologically identifiable variables. A given set of principles can have more than one possible physiological embodiment. For example, both Booth (1978) and Russek (1978) have proposed models of feeding control which are based upon negative feedback principles with modifications to incorporate the influence of circadian rhythms, etc. However, these models are very different in their physiological implications. Booth's ischymetric model is based upon rates of energy metabolism, whereas Russek's model depends upon hepatic glucoreception.

5.1 Classical control systems theory

The rigorous quantification of motivational variables owes much to the application of classical control systems theory, both at the physiological level (e.g. Bray *et al.*, 1978) and at the purely behavioural level (e.g. McFarland, 1971, 1978a; Toates, 1975). Traditional psychological measures of motivational state include the various indices of the amount of work that an animal is prepared to do to obtain certain rewards, the amount of electric current an animal is prepared to endure to obtain a reward, the latency or speed of performing certain tasks, etc. (Bolles, 1975). Such methods rarely give reliable quantitative indices of motivational state, either because they are usually contaminated by interactions with other aspects of behaviour, such as avoidance, or because they are unsatisfactory from the point of view of measurement theory (see sections 4.2 and 4.3). The advantage of the classical control systems approach is that it is capable of taking account of interactions between different systems, while at the same time it is open to a holistic type of verification. Any control-model, or computer simulation, is an hypothesis which produces complex predictions that can be tested empirically. It is possible to test whether the theory works as a whole, whereas traditional psychological theories have to be tested in a piecemeal fashion.

5.1.1 Classical control models

Classical control models portray hypotheses in the language of control systems theory, which is often represented in the form of a block diagram or flow graph. The advantages and disadvantages of this convention have been discussed at length elsewhere (e.g. McFarland, 1971, Toates, 1975).

There are generally two stages in the development of a classical control model. In the first stage a provisional model is formulated by one means or another, and in the second stage the model is tested for stability, for robustness, for its ability to make useful predictions, etc. Once a model has been developed it can be tested experimentally in the normal scientific manner. A number of approaches can be used in the development of a provisional model. A simple model may suggest itself on purely logical grounds (e.g. McFarland, 1965b), or on physiological grounds (e.g. Oatley, 1967). A model may be developed from a more simple model by the process of temporal dissection, as advocated by McFarland (1971). By making an initial study on the basis of large time-units, it is possible to arrive at a model with few large time-constants, because the quick-acting processes do not reveal themselves when the time-unit is large. A relatively simple framework-model can be achieved in this way, and this can be dissected into its more complex components by progressive shortening of the time-unit, which reveals those aspects of the system which have shorter time-constants.

Another useful approach is stochastic analysis of observed data. For example, as a result of sophisticated stochastic analysis of aggressive behaviour of the cichlid fish (*Haplochromis burtoni*) Heiligenberg (1974) arrived at the control model illustrated in Fig. 5.1. Various types of stochastic analysis have been successfully employed in the

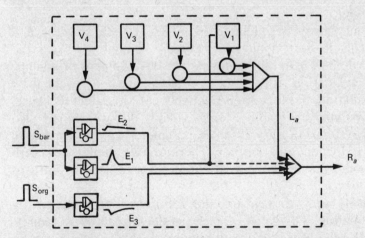

Fig. 5.1 Analog model representing readiness to attack (R_a) in a cichlid fish. R_a is the sum of an internal variable (L_a) and of processes (E) which are triggered by an external stimulus S. $V_1 - V_4$ are the internal states obtained by stochastic analysis. (After Heiligenberg, 1974.)

formulation of motivational models, including multivariate analysis of variance (Pimental and Frey, 1978); hierarchical cluster analysis (Dawkins, 1976; de Ghett, 1978); Markov chain analysis (Metz, 1974) and stochastic identification techniques (Lloyd, 1974, 1975).

Once a provisional model has been formulated it must be evaluated. The generally accepted method of evaluating a model is by its predictions. It is important, however, to attribute to each prediction, with its success or failure, a judgement in relation to the assumptions upon which the prediction is based (Dawkins and Dawkins, 1974). Disproof of predictions may not invalidate the whole model, but only certain assumptions. Certain evidence, on the other hand, may be sufficient to disprove a whole class of models (see Doucet and van Straalen, 1980). For example, there are many models of feeding and drinking behaviour in which the satiation process is represented as a negative feedback system. This generally results in an exponential satiation curve that conforms well with the observed phenomena. However, there is some evidence that animals 'track' the satiation curve, rejoining the curve if experimentally forced off it (McCleery, 1977). McFarland (1971) has suggested that satiation is sometimes achieved on a pre-programmed basis, and not on the basis of feedback at all. Nevertheless, the feedback model fulfills many predictions and remains the basis of most current efforts to simulate the satiation process. This raises the question of whether the confirmation of predictions is necessarily a good guide to the value of a model.

Models should also be evaluated in terms of their stability and robustness. This is best done by computer simulation, since the complexity of models generally outpaces the development of analytical means of evaluation. Instability in a model is soon revealed by computer simulation and it is often relatively easy to detect its cause. That living systems are stable is self-evident, but we should also expect living systems to vary in their parameter values, especially between one individual and another. Therefore we should view with suspicion a model that is stable only within a set of narrow limits of parameter values.

It is relatively easy to test for the robustness of a model by computer simulation. The value of each parameter can be systematically varied, and the sensitivity of the output to this variation can be measured. Robust models are relatively insensitive to parameter variation, and this is generally considered to be a desirable feature of a model.

If a model is to be more than a redescription of observed data, it must contain an element of hypothesis that can be empirically tested. There will often be many possible hypotheses capable of accounting for a given set of data, and some systematic exploration of alternative possible models is often a good policy. Geertsema and Reddingins, (1974), in their simulation of feeding behaviour, and Houston, Halliday and McFarland (1977) in their simulation of newt courtship, provide good examples of overt exploration of different possible models.

Classical control systems analysis has proved to be particularly well suited to the type of problem in which the specific dynamic relationship between measurable variables is to be investigated. Such a relationship can often be described in terms of a linear transfer function or in terms of a straightforward block diagram. There are many well established empirical methods for determining transfer functions, such as transient analysis, frequency analysis, random input analysis, etc. (McFarland, 1971; Toates, 1975). These approaches have been used successfully in biological systems analysis (McFarland and Budgell, 1970; Oatley and Toates, 1971). Classical control systems theory has also proved useful in the synthesis of complex subsystems, such as are involved in respiratory and cardiac physiology. Computer simulation of physiological processes is now a major means of gaining an holistic understanding of complex physiological systems, and this approach has been extensively applied in the study of the control of feeding and drinking (e.g. Booth, 1978).

5.1.2 Homeostasis

The concept of homeostasis has played a dominant role in the psychology of motivation, and it was important in the early applications of control systems theory. Although it is widely recognized (e.g. McFarland and Nunez, 1978) that too much emphasis has been given to the role of homeostasis, it nevertheless plays an important part in current motivational thinking.

McFarland (1970) justified the application of control theory to behaviour by reference to the stability of the internal environment as described by Claude Bernard (see Chapter 2). Bernard's ideas were developed by Cannon (1932) to produce the concept of homeostasis, which has been defined as follows:

'Homeostasis in its widest context includes the co-ordinated physiological processes which maintain most of the steady states in organisms.... It must be emphasised that homeostasis does not necessarily imply a lack of change, because the 'steady states' to which the regulatory mechanisms are directed may shift with time. But throughout the change they remain under more or less close control.' (Preface to the 18th Symposium of the Society for Experimental Biology, 1964.)

The above definition of homeostasis indicates that physiological states are not kept constant, but rather are regulated so as to remain within certain bounds. A weakness in the definition is that it makes no mention of the role of behaviour in the regulation of physiological state. McFarland (1965b) and Oatley (1967) made this role explicit by incorporating drinking into models for the control of water balance. However, such models generally involve the simplification of considering one physiological variable that is maintained at a normal or 'optimal' value. This type of approach leads to simple negative feedback models in which a discrepancy between a desired value and an actual value results in behaviour, the integrated consequences of which remove the discrepancy.

However, recent research (McFarland, 1970, 1978a; Toates, 1979) shows that to account for the control of such classical homeostatic aspects of behaviour as feeding, drinking and thermoregulation, it is necessary to consider not only the negative feedback mechanisms involved, but also feedforward anticipatory processes, adaptive control mechanisms involving learned changes in behavioural strategy, and acclimatory mechanisms providing long term physiological adjustment. As McFarland and Nunez (1978) point out '... we can see no justifiable basis for distinguishing between homeostatic and non-homeostatic forms of behaviour control. The maintenance of the internal environment is not simply a matter of simple negative feedback processes, but of complex interactions between feedback, anticipatory, adaptive and optimizing control. In these respects the control of sexual behaviour is not fundamentally different from the control of feeding or drinking' (p. 626). In the next section we illustrate this point.

Although classical control theory, and its counterpart in computer simulation, can be applied to specific and intricate problems of behaviour control with considerable success (Booth, 1978; McFarland, 1971, 1974a; Toates, 1975), it is not suitable for handling problems involving interactions between many different types of behaviour; for these a state-space approach is much better suited (McFarland, 1978a). Nevertheless, classical control models can be used to illustrate certain fundamental concepts in model making. In section 5.2 we introduce a model which will enable us to discuss these issues.

5.1.3 Classical control in relation to the state–space approach

Classical control models can relatively easily be translated into the state–space format, once the state–space variables have been identified. A linear system can be described in terms of a transfer function in which a number of state variables can be identified.

By definition, a transfer function is the Laplace transform of a differential equation with zero initial conditions. The order of the differential equation is the same as the number of state variables, and the number of initial conditions that must be specified in order to predict the future behaviour of the system. Representation of linear systems in state-variable form involves n first-order differential equations, as opposed to the usual nth order equation (Schultz and Melsa, 1967). Thus the first step in deriving a state-variable representation from a transfer function is to obtain the n first-order differential equations. For example, the transfer function

$$H(s) = \frac{K}{s^3 + k_1 s^2 + k_2 s + k_3} \tag{5.1}$$

is equivalent to the following third-order differential equation,

$$\frac{d^3 y}{dt^3} + k_1 \frac{d^2 y}{dt^2} + k_2 \frac{dy}{dt} + k_3 y = Kx(t) \tag{5.2}$$

Let

$$y(t) = a \quad \frac{dy}{dt} = b \quad \frac{d^2 y}{dt^2} = c \tag{5.3}$$

then

$$\frac{dc}{dt} + k_1 c + k_2 b + k_3 a = Kx \tag{5.4}$$

giving the three first-order differential equations,

$$\frac{da}{dt} = b \tag{5.5}$$

$$\frac{db}{dt} = c \tag{5.6}$$

$$\frac{dc}{dt} = Kx - k_1 c - k_2 b - k_3 a \tag{5.7}$$

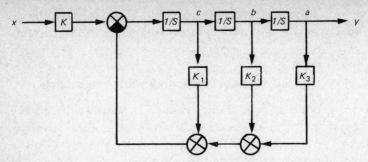

Fig. 5.2 Phase-variable representation of a third-order system. (From McFarland, 1971.)

The variables b and c, being derivatives of the system output a, are known as *phase variables*. In general phase variables are defined as those state variables which are obtained from one of the system variables and its $n-1$ derivatives (Schultz and Melsa, 1967). The variables most commonly used are the system output and its derivatives. The phase variable representation of a system can be portrayed in block diagram form, as illustrated in Fig. 5.2. Phase variables do not generally correspond to physical variables that can be measured or manipulated experimentally. Although phase variables are useful in certain types of analysis, particularly in describing behaviour of certain types of non-linear systems in terms of the phase plane, they do not enable the investigator to take equal account of all the important state variables in a system. A more intuitive approach is to consider those system variables that are physically meaningful.

A system transfer function can generally be broken up into a number of subsystems, each containing a single integrator. There will generally be a number of different ways in which this can be done, as illustrated in Fig. 5.3. It is not possible to say whether the

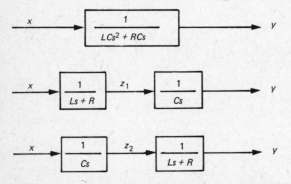

Fig. 5.3 Alternative ways of dividing a second-order system into two first-order subsystems. (After McFarland, 1971.)

state variables z_1 and z_2 are physically meaningful, without further knowledge about the system. If the complete block diagram of the system is known, then it is a simple matter to choose physically meaningful state variables. For example, Heiligenberg's model for aggression in cichlid fish, illustrated in Fig. 5.1, has an internal variable L_a which corresponds to the sum of various internal processes $v_1, \ldots, v_4$. As Heiligenberg (1974) points out:

> 'The correlation analysis of slow fluctuations in the rate of different behaviours led to the conclusion that at least four basic slow processes, v, are involved in the causation of the seven behaviours investigated. The long-term fluctuation in the rate of each activity could be represented by a vector in a four-dimensional space subtended by four orthogonal processes, $v_1, \ldots, v_4$, and the projection of this vector onto each v-coordinate taken as the proportion the particular v-process takes in the control of this behaviour. Unfortunately, the representation of the seven activities within a given four-dimensional v-space as shown in Table 2.1 and Fig. 2.8 is not unique, because any orthogonal transformation of the seven vectors within their four-dimensional space would not affect the correlation pattern found. This is equivalent to saying that the activities investigated can be represented by vectors within a four-dimensional space without any particular v-coordinate system being specified. One may thus ask whether any slow processes could be identified on a physiological level in order to specify the long-term processes, v, and to arrive at a unique representation of these processes within the seven activities investigated. For this purpose experiments must be performed to evoke long-term changes in the rate of these activities. If, for example, the application of a certain hormone, such as testosterone, caused a long-term rise in the rate of activities 2, 3, 4 and 5 (Figs. 2.5 and 2.8) one would assume this hormone to affect a process comparable to v_1 (Fig. 2.8). The proportion to which these four activities are affected would then yield a particular representation of this hypothetical long-term process within the four activities concerned. The application of a different hormone may affect a different set of activities and lead to the determination of another slow process, v.' (pp. 107–8).

5.2 A model of the courtship of newts

The behaviour to be modelled is the courtship of the male smooth newt, *Triturus vulgaris*, as described by Halliday (1975, 1976). The courtship takes place in the spring when the newts return to ponds after hibernating on land. During this aquatic phase, newts must swim up to the pond surface to breathe, so they must hold their breath during the under water courtship. Fertilization is internal and is achieved by the male depositing a spermatophore on the floor of the pond. The female than walks over the spermatophore and, in a successful courtship, picks it up with her cloaca. The model we wish to discuss is only concerned with the two activities of the male that precede this transfer and which Halliday calls retreat display and creep. During retreat display, the male backs away from the advancing female while performing various movements with his tail. This phase ends when the male turns away from the female and creeps along the floor of the pond. The female follows him and when he quivers she touches his tail with her snout. This appears to be the stimulus for him to deposit a spermatophore. Halliday (1975) found that on the first of several consecutive spermatophore depositions, males tended to complete this prototypical sequence (retreat display-creep-spermatophore transfer) without interruption. On the last spermatophore however, they often alternated between retreat display and creep before depositing a spermatophore (see Fig. 5.4). Furthermore, the display rate is correlated with the number of spermatophores still to be deposited. This number, which is known only at the end of the courtship, will be referred to as S (S usually ranges from 1 to 3).

We have described elsewhere (Houston, et al., 1977) the development of a model of the male's behaviour. The final version, called

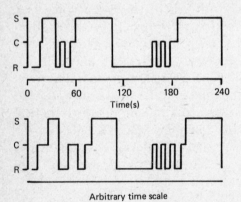

Fig. 5.4 Comparison of mean durations of various courtship activities of 22 newts (above) with that of the NEWTSEX model (below). R = retreat display, C = creep, S = spermatophore transfer. (After Houston et al., 1977.)

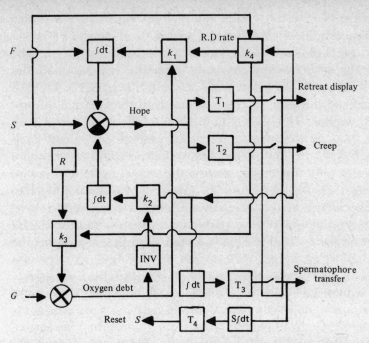

Fig. 5.5 A control model for the courtship of the male smooth newt. The following conventions have been observed: S = initial amount of sperm, F = initial sexual state of the female, G = initial value of oxygen need, R = function representing the increase in oxygen need with time, $T_1 - T_4$ = thresholds, $k_1 - k_4$ = modifiable parameters. (After McFarland, 1977, modified from Houston et al., 1977.)

NEWTSEX IV, is shown in Fig. 5.5. The model is based on a variable called hope, which can be thought of as the male's assessment of the sexual state of the female. The initial value of hope depends on S, and its rate of change is determined by the consequences of the male's behaviour, represented by feedback loops with parameters k_1 and k_2. During retreat display, when the male can see the female approching him (represented by F in Fig. 5.5) hope rises (i.e. the feedback is positive). When the male turns away to creep he can no longer see the female, so it is assumed that hope falls because of negative feedback. The transition from retreat display to creep is determined by a threshold called T_2. Once creep begins, hope starts to fall, so that if T_2 were also the threshold for the transition from creep to retreat display, the model would get trapped on it. In NEWTSEX IV this difficulty is overcome by having a lower threshold, T_1, which determines the transition from creep to retreat display. This means, however, that between T_1 and T_2 the value of hope does not suffice to determine whether retreat display or creep occurs: the direction in which hope is changing must also be known.

To explain the influence of the number of spermatophores to be deposited on the nature of the courtship sequence, it is assumed that the rate at which hope falls is inversely proportional to S. NEWTSEX IV assumes a fixed duration of creep after which the spermatophore-transfer phase begins. This means that if hope falls slowly, as it does when $S = 3$, there is time for creep to be completed before T_1 is reached. When $S = 1$ or 2 hope may get to the threshold before there has been enough time for creep, and so the model switches back to retreat display (see Fig. 5.6). As it stands, if this model fails to complete creep at the first attempt, it will alternate between retreat display and creep indefinitely. To enable it to progress to the transfer phase, we introduce a variable called oxygen need that represents the internal availability of oxygen. We do not want to make any precise statement about the physiological basis of this variable; we merely assume that it increases with time under water, and that the rate of increase is proportional to the newt's rate of activity. This increase is represented by the ramp function R which is influenced by the rate of retreat display. The size of the gulp of air the newt takes at the surface is modelled by the value of G, which is subtracted from the ramp function.

Adding the oxygen need variable to the model enables the alternation between retreat display and creep to be discontinued because it

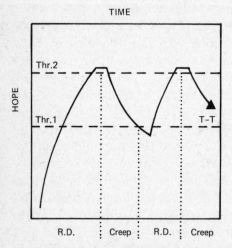

Fig. 5.6 The behaviour of the hope variable in the NEWTSEX model. T–T = tail touch. (After Houston *et al.*, 1977.)

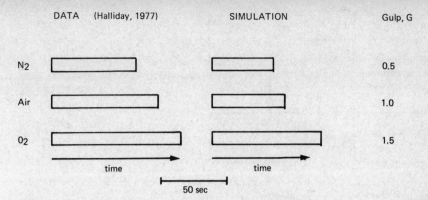

Fig. 5.7 Duration of newt courtship from the start of retreat display to the deposition of the first of three spermatophores, under three conditions: high nitrogen content, normal air, high oxygen content of water and surface gases.

is assumed that as oxygen need increases, hope falls less rapidly during creep. As can be seen in Fig. 5.4 the output of the resulting model is similar to the behaviour of the male newts. Some implications of the model have been tested by Halliday (1977). He changed the atmosphere above the water in the newts' tank, which should change the value of G and hence the timing of the courtship sequence. If it is assumed that the value of G is bigger than usual when the atmosphere is enriched with oxygen and smaller than usual when extra nitrogen is added, then the model predicts that the courtship will be slower than normal in the former case and faster in the latter. This prediction was supported by the results, as is shown in Fig. 5.7.

5.3 The consequences of behaviour

NEWTSEX IV illustrates that the framework of control theory can be used to model behaviour even when there is no physiological variable that is being regulated. Models similar to NEWTSEX IV have been developed for the sexual behaviour of the male rat (Freeman and McFarland, 1974; Toates and O'Rourke, 1978), and a review of these and other examples is given by Toates and Archer (1978). The important general concept in these models is that they emphasize feedback rather than simple regulation. On the basis of this and our discussion of causal factors, we suggest the following

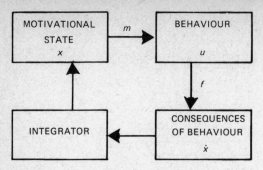

Fig. 5.8 The closed-loop linking motivational state *x* and behaviour *u*. m = motor parameters, f = feedback parameters.

framework for the modelling of behaviour sequences:

(1) The activity that an animal performs is uniquely determined by a set of causal factors.

(2) The way in which the causal factors change depends on the consequences of the activity that the animal is performing.

This formulation results in a closed loop linking motivational state and behaviour, as is shown in Fig. 5.8. (This is not to say that all the consequences of an animal's behaviour are fed back to change an animal's state, or that such changes depend only on the consequence of the animal's behaviour.)

Models of this sort can be represented by a 'map' defined on causal factor space such that within each region of the map just one activity occurs, and this activity determines the movement of the state in causal factor space. This idea permits a general representation of behaviour as inducing a 'flow' on state space (Houston, unpublished). An introduction to this way of looking at dynamic systems can be found in Rosen (1970). As has already been pointed out, NEWTSEX IV does not fall into this category, because between T_1 and T_2 the values of hope do not specify which activity is performed. Houston (1977) describes a variant of NEWTSEX IV in which there is a unique relationship between the variables and the activities. This is achieved by introducing a variable which represents positive feedback from creep. There is then no need for a pair of thresholds on the hope variable, and the map shown in Fig. 5.9 results.

When attempting to assess such models, the following questions would seem to be pertinent: Does the model contain any redundant elements? How many other models could also account for the data? In the next section we show that these questions are closely related.

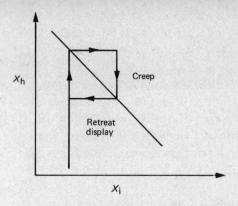

Fig. 5.9 Map of Houston's (1977) variant of NEWTSEX IV. x_h = the primary causal factor, x_i = the positive feedback from creep behaviours.

5.4 The concept of observability

The relationship between the transfer function and the state space representations was established by Kalman (1963). He introduced the concepts of controllability and observability, which have become central to system theory. A system is completely controllable if it is possible to move any state to any other state in a finite time by a suitable choice of input. A system is completely observable if an observation of the inputs and outputs for a finite time enables the state to be determined. Gilbert (1963) has shown that systems can be partitioned into subsystems that are controllable and observable, controllable but unobservable, etc. (Of course, not all systems will have all possible subsystems.) This idea is represented in Fig. 5.10

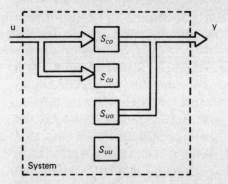

Fig. 5.10 Partitioning of a system on the basis of controllability and observability criteria. Thick arrows indicate matrix relationship. (After Schultz and Melsa, 1967.)

which illustrates that uncontrollable states are detached from the inputs and unobservable states are not represented in the outputs. Kalman showed that, for linear systems, the transfer function is equivalent to those states of the system that are controllable and observable. This result has two implications. (1) If a technique for obtaining a transfer function is applied to a system, any uncontrollable or unobservable parts of the system will not be represented. (2) If a completely controllable and observable model is constructed, then a behaviourist has no grounds for objecting to its state variables, for they are equivalent to the input-output representation of the transfer function.

Turning to the questions that closed the previous section, we introduce the following terminology. If a system contains no redundant elements, it is said to be minimal; if a system is the only one that accounts for the data (ignoring systems that are the same except for the names of the states) it is said to be unique. It is obvious that if we consider unobservable systems, we cannot obtain minimal or unique models, because the models will contain elements that do not contribute to the output. It turns out that controllability is also essential (this may not be so obvious, but in fact controllability is the mathematical 'dual' of observability (Schultz and Melsa, 1967)) as was shown by Kalman (1963) who established the following results: A model is minimal if and only if it is completely controllable and completely observable. A minimal model is unique. A summary of these and related results can be found in Kalman (1968).

McFarland (1971) took up these ideas and suggested that they were of relevance to the study of behaviour. One example he gave was that of song-learning in birds. The young bird may be thought to have a song 'template', which, in some species, can be modified by the song the bird hears before it starts to sing. McFarland argued that, during the sensitive period, the template is controllable but unobservable.

A more detailed application of the concept of observability is given by McFarland and Sibly (1972). They consider the implications of an environment in which an animal is unable to reach a given point in state space (see section 2.3). To be specific, imagine that two axes of the animal's state space are fat deficit and protein deficit and that there are two sorts of food, red and green. Green food contains 50% fat and 50% protein, whereas red food contains 90% protein and 10% fat. The consequences of eating red or green food can be represented as vectors in a fat-protein consequence space, as is shown in Fig. 2.9. From any point between vectors **R** and **G** it is possible to reach the origin by eating a mixture of the two foods; such points are said to be realizable. Given an unrealizable point, such as x_p in Fig. 2.10, it is assumed that the animal brings its state as near to the origin as possible. Specifying the measure of nearness to the origin amounts to stating the optimality criteria by which the animal weights the deficits on the various axes. Rather than make any unwarranted assumptions about this criterion, we may hypothesize that the animal will behave in the same way for various initial deficits when attempting to move its state as near to the origin as possible. This is illustrated in Fig. 5.11. Here all the points on the line PQR result in identical trajectories in state space, brought about by identical consumption of red food. This means that the state is unobservable. McFarland and Sibly's statement that observable commands are in one-to-one correspondence with their optimal consequences might seem to introduce a slightly different notion of observability from the one already presented, but in fact it is equivalent to the definition of observability used by Arbib and Zeiger (1969) (see the next section for further details). An application of the more usual definitions to a physiological system is described by Kapur, Ghosh and Nath (1976).

Although Kalman's work has done much to clarify the principles of model-making, it must be recalled that his results apply only to linear systems. In the next section we consider non-linear systems, for which there are no general results.

5.5 Non-linear systems

A system is said to be linear if its output is a linear function of the input. For a system with one input variable u and one output variable y, this means that

$$y = au + b$$

where a and b are constants.

For systems with more than one input or output, engineers often define a linear system as one that obeys the principle of superposition (e.g. Elgerd, 1967). This principle states that if Y_1 and Y_2 are the outputs produced by the inputs U_1 and U_2 respectively, then the output produced by $U_1 + U_2$ is equal to $Y_1 + Y_2$. In the context of transfer functions, a system is linear if the transfer function does not depend on the amplitude of the input.

Biological models often involve non-linearities (Clynes, 1961; Oatley, 1967) and the NEWTSEX IV model is no exception. Given that it is not possible to obtain a complete theory of non-linear systems, one line of attack is to consider individual cases. For example, Norman and Gallistel (1978) consider the problem of finding the tranfer function of a linear time-invariant system whose output can only be seen if it is greater than a certain threshold value. This means that the whole system (i.e. the linear system plus the threshold) is non-linear, but Norman and Gallistel show that, under mild restrictions, the linear part can be determined by the use of various increasing or decreasing inputs.

An alternative approach involves abandoning some of the system's structure and treating it as a finite automaton. This might seem to be a drastic change of view, but there are close correspondences between automata theory and dynamic systems theory. Kalman (1963, especially pp. 169–170) compared his theorems with Moore's (1956) results on automata and Arbib (1966, 1969) has investigated this relationship in considerable detail.

A finite automaton (or machine) M consists of a finite set of inputs, $\underline{U}$; a finite set of states, $\underline{X}$; a finite set of outputs, $\underline{Y}$; a next-state function δ and an output function λ. The machine operates on a discrete time scale; δ specifies the state the machine moves to on the next time unit, given the current state and input, while λ describes the relationship between the current output and the state and input at the previous time unit.

Automata theory has two advantages over the linear systems approach: the states can be established in a neat fashion, and the question of linearity does not arise unless extra structure is added. The states are defined by a concept called Nerode equivalence. This is an equivalence relation between pairs of input strings. For a pair of strings Q_1 and Q_2, consider two identical copies of the machine in question and apply Q_1 to one machine and Q_2 to the other. If the response to any further input is the same for each machine, then Q_1 and Q_2 are Nerode equivalent. The machine's states are identified with the sets of Nerode equivalent states, so that a state can be thought of as a set of equivalent histories. To see that this definition of states agrees with the one we have previously given, note that for any member of a Nerode equivalence class, the output depends only on the input, i.e. specifying the equivalence class and the subsequent input suffices to determine the future output.

This approach provides a representation of the input-output data in which every state can be reached from one initial state and every state has a different behaviour. These properties, which can be thought of as corresponding to controllability and observability respectively, imply that the model is unique up to the relabelling of its states (see Arbib 1969, for further details). Once such a machine has been obtained, it may be possible to give its states a topology and produce a continuous-time model. An elegant example of this is given in Metz (1977), where Nelson's (1965) model of the behaviour of the male stickleback is re-interpreted.

We can justify the claim that the automata-theory notion of observability, in which no two states have the same output for all inputs, corresponds to that used by McFarland and Sibly (1972). The argument is based on identifying consequences with the output of a machine, from which it follows that the states Q and R in Fig. 5.11 have the same output and are therefore unobservable. As McFarland and Sibly (1972) put it, 'observable commands are in one to one correspondence with their optimal consequences'.

Our presentation of Nerode equivalence has involved certain simplifications; for instance an infinite number of experiments on the system in its initial state may be required to establish the equivalence classes. There is, however, the following useful result: if at any stage the length of the input strings is increased by one input and no further increase in the number of equivalence classes occurs, then all the states have been discovered. Further details can be found in Arbib and Zeiger (1969, section 3).

Although the concept of observability has done much to improve

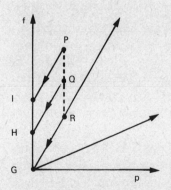

Fig. 5.11 Command space showing three trajectories which have the same behavioural consequences. They are therefore unobservable. (After McFarland and Sibly, 1972.)

our understanding of model-making, the theory rests on the assumption that, the way the state changes is known. Within our framework, this change is, to a large extent, determined by feedback from the consequences of the animal's behaviour. McFarland and Sibly (1972) draw attention to the importance of this feedback in establishing observability and the arguments from automata theory have substantiated this point. Furthermore, the closed loop between behaviour and motivational state that results from this feedback forms the basis for the application of optimality theory, as we show in Chapter 8.

5.6 Summary

The relevance of classical control theory to quantitative ethology is discussed, with special emphasis upon the ways in which models can be evaluated. Shortcomings of classical methods can be overcome by the state–space approach.

Models of the courtship of the male smooth newt are used to illustrate some of the problems inherent in quantitative ethology, especially problems of observability, controllability and non-linearity.

6 Optimal decision-making

The survival and reproductive success of an individual animal depends largely upon the animal's use of resources, such as food, territory, mates, etc. At any particular time, an animal may have alternative courses of possible action so that a choice has to be made. Every activity will have associated costs and benefits in terms of the ultimate reproductive success or fitness of the animal. For example, a male stickleback, which moves from his nest to court a female, may be increasing his chances of successfully luring the female to his nest and inducing her to deposit her eggs there, but he may be endangering his nest by neglecting to guard it against marauding neighbours, or to repair it after inadvertent disturbance by other animals (Wooton, 1971; McFarland, 1974b). Thus there are both benefits and costs associated with the courting activity. In general, the costs and benefits are attached to both the behaviour of the animal and its internal state. Suppose we consider the situation of a deer in cold, wet weather. The animal has a choice between standing up to feed and sitting down to shelter from the wind. There are costs and benefits directly associated with the behaviour. Thus by standing the animal has a good field of view and can easily look out for possible danger. It can feed, which it cannot do effectively while sitting, but it is much more exposed to the weather than it would be sitting in a sheltered place. By sitting the animal is able to save energy, both by being less active and by reducing heat loss in sheltering from the cold wind. On the other hand, it is less well placed to spot possible predators or rivals, and it cannot easily feed. Clearly, in assessing the relative merits of the two possible behaviours, the animal's state of hunger should be taken into account. If the animal is not very hungry, it can probably afford to sit down and wait for the bad weather to pass. If on the other hand it is in need of food, it may endanger its life by neglecting to feed now, since it may have to flee from a hunter or other danger at some time in the future when it would otherwise have fed. Even for such a simple decision as sitting down or standing up, the balance of costs and benefits may be a delicate one.

This much is biological common sense, but it raises a number of difficult issues and queries. Can the costs and benefits be quantified? Are there any general principles operating within this aspect of animal decision-making? Does the cost–benefit analysis envisaged in animal behaviour bear any relationship to economics?

6.1 Functional aspects of decision-making

6.1.1 Analogies between economics and ethology

A general principle, subscribed to by both economists and ethologists, is that in the processes of decision-making something is maximized. At the global level we can see that whereas biologists account for behaviour in terms of genes or fitness, the economists or sociologists have to resort to the less well defined concept of value, as ilustrated in Fig. 6.1. Whereas fitness may be clearly defined in terms of genes and some phenomena may be said to impart greater fitness than others, value is less clearly defined in terms of prevailing ideas. Some phenomena are said to have greater social value than others, although what is of social value may be a matter of opinion.

At the level of the individual, we have a more direct parallel between the concepts of *cost* and of *utility*. Cost is a notional measure of the contribution to fitness of the behaviour of an animal and of its

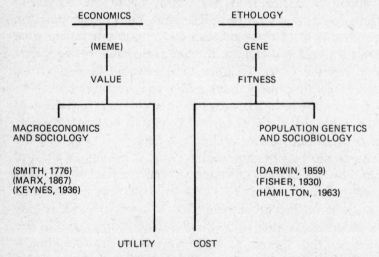

Fig. 6.1 Parallel concepts in economics and ethology. The names in parentheses are those of the most important contributors (with the dates of their major works) to macroeconomics and population biology.

motivational state (Sibly and McFarland, 1976). Utility is a notional measure of the psychological value of goods, leisure, etc. Thus I may obtain a certain amount of personal satisfaction or utility from attending a conference, from sport, from eating, etc. For an animal building a nest, the utility of nest material would generally be a decelerating function of the amount of nest material already obtained (Fig. 6.2(a)). In biological terms we might say that the cost of not having nest material declines with increasing amount of nest material (Fig. 6.2(b)). Economists would call this decreasing marginal utility. The implication here is that a component of the risk to reproductive success is inversely related to the amount of nest material gathered. Thus we can see that the concepts of cost and utility are the inverse of each other, and we can speak of animal as maximizing utility or as minimizing cost.

In economics it is taken as axiomatic that the rational economic man is a maximizing agent, and the function that is maximized is

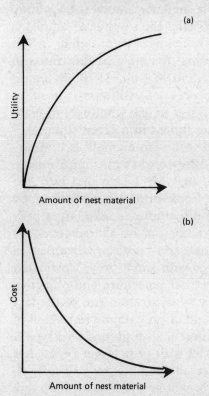

Fig. 6.2 (a) The decreasing marginal utility of nest material, and (b) its equivalent cost.

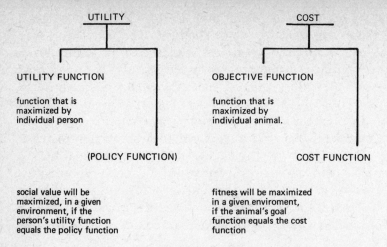

Fig. 6.3 Parallel concepts in microeconomics and ethology.

generally called the *utility function*. It is similarly argued (McFarland, 1977; section 6.2.1) on the basis of decision-theory that an animal is also an optimizing machine, and the function corresponding to the utility function can be called the *objective* function, as illustrated in Fig. 6.3. Many biologists (e.g., Oster and Wilson, 1978; Maynard Smith, 1978) take the view that the process of evolution is itself an optimizing process and that it makes sense to ask what evolutionary design or strategy is likely to maximize fitness in a given situation.

Although natural selection will tend to design animals as optimizing machines, this does not mean that they always maximize fitness. Suppose we imagine a species of bird that can be divided into two varieties on the basis of feeding habits. Some birds are specialized, by virtue of morphological and behavioural features, to take larger prey than others. Suppose that the two varieties can be characterized by beak length so that the bird with a larger beak is at an advantage in taking large prey but at a disadvantage with small prey. Correlated with beak length is a particular behavioural repertoire and objective function, which are characteristic of the specialization and of the fact that the animal searches for the right kind of prey, in the right kind of place, etc. On a graph of prey size plotted against fitness, bird type α does best with prey size A, and less well with prey larger or smaller than A. Similarly bird type β is best adapted to prey size B, as illustrated in Fig. 6.4(a). Normally there will be niche separation due to competition between the two varieties. Suppose, however, that bird β is obliged to take prey A due to scarcity of prey B in the

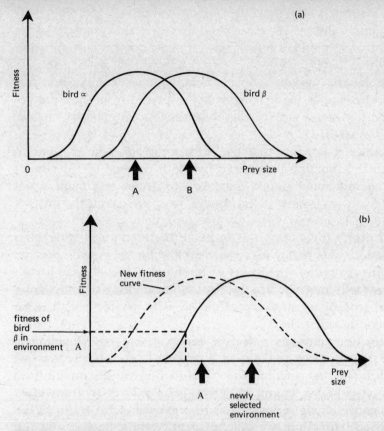

Fig. 6.4 (a) Graph of prey size against fitness for two types of bird, α and β. α is adapted to prey size A, and β is adapted to prey size B. (b) Bird β in environment A may increase fitness by moving to a new environment, or by modifying its objective function to suit environment A (dotted curve).

environment. Clearly bird β will have reduced fitness, as indicated in Figure 6.4(b). If we take prey size as an index of the type of environment, we can see that bird β is best adapted to environment B, and we can expect it to have an objective function designed to suit environment B. However, we can imagine that there is a possible objective function that could be designed to suit environment A, so that an animal having this objective function (such as bird α) would be best adapted to environment A. In other words we can distinguish between the function that an individual actually maximizes – called the objective function – and the function that it should maximize if it were perfectly adapted to the environment. This latter function is called the *cost function* (McFarland, 1977; McCleery, 1978), and it is essentially a property of the environment rather than of the individual animal. Only when an animal is perfectly adapted to its environment is its objective function identical to the cost function relevant to that environment. In reality, this will rarely be the case because of genetic variation between individuals, competition between individuals resulting in some being displaced from their preferred habitat, and evolutionary lag following environmental change (see also Chapter 9).

In economics it appears that the term utility function is used to denote both the function maximized by the individual and the function that the individual should maximize to attain maximum social value in a given environment. In Fig. 6.3 we have called the latter a *policy function*. In everyday life we are continually under pressure to change our utility functions so as to make them conform with some currently fashionable policy function. As children we may be pressed by our parents to reduce the utility of bubblegum and replace it with something socially more acceptable. As adults we may be pressed by government to give up smoking or gin so as to conform with some current public health policy.

The distinction between objective function and cost function is reflected in two distinct methods of approach to the adaptedness of behaviour. In the case of the objective function we attempt to find what function is being optimized by the behaviour of the individual animal. Examples of this approach are discussed in Chapter 9. In the case of the cost function, we aim to identify those features of the environment which determine the costs and benefits of the different aspects of the behaviour of the animal. Such studies normally involve field work and constitute the major part of the current effort in behavioural ecology.

This type of work includes attempts to specify the costs of foraging in relation to the benefits gained (e.g., Davies, 1977; De Benedictis *et al.*, 1978; Goss-Custard, 1977; Krebs, 1978; Werner and Hall, 1974), the costs and benefits of territorial behaviour (Carpenter and MacMillen, 1976; Davies, 1978; Davies and Houston, 1981), the costs and benefits of thermoregulation (Huey and Slatkin, 1976).

6.1.2 The animal as a consumer

In recent years there have been a number of hypothetical models of foraging behaviour in animals which are based upon the assumption that the strategies employed by predators are designed to maximize the net rate of energy intake (Charnov, 1976a, b; Emlen, 1966; MacArthur and Pianka, 1966; Schoener, 1971; Pulliam 1974; Werner and Hall, 1974). In many of these studies there is also an explicit or implicit assumption that energy or some other component of food can be taken as a convenient metric of foraging success. At the theoretical level this approach is open to the criticism that to study the costs and benefits of foraging in isolation from other aspects of the animal's life is unrealistic. Thus, it may be that the most profitable means of foraging exposes the animal itself to predation, but that a less profitable strategy is not so exposed to such dangers. How should the animal weigh one strategy against the other? Some authors (e.g., Krebs, 1973) recognize that fitness is the only ultimately satisfactory criterion by which the relative merits of different aspects of behaviour can be judged, but regard net energy balance as the most convenient metric in the case of foraging behaviour. The question is, therefore, to what extent do empirical studies justify this approach? From a review of the literature (e.g., Krebs, 1978), it is clear that animals do, under some circumstances, appear to maximize energy profitability. However, it is also clear that in other circumstances this is not the case.

This point is well illustrated by Goss-Custard's work on foraging by redshank *Tringa totanus*. In a field study, Goss-Custard (1977) found that redshank foraging on the shore preferred the size classes of their prey, the polychaete worms (*Nereis diversicolor* and *Nephthys hombergi*), which had the highest return per unit handling time. This preference for large prey was dependent on the encounter rate with large, but not with small prey. However, Goss-Custard (1977) also found that when the amphipod *Corophium* was available in addition to polychaete worms, the birds preferred *Corophium*. Redshank were increasingly unlikely to take a worm that they encountered as the density of *Corophium* increased. The possibility that the habitat typical of *Corophium* was one in which the worms were harder to find was discounted by the observation that some individual redshank take worms when the majority are feeding on *Corophium*. Goss-Custard formed the hypothesis, derived from a number of discussions on prey selection (Emlen, 1966; MacArthur and Pianka, 1966; Schoener, 1971; Pulliam, 1974) that by preferring *Corophium* redshank achieved a higher rate of intake of energy than they would have attained by taking worms. This hypothesis was concordant with the widely held belief that, in contrast to many herbivores, carnivores are more likely to select for energy than for scarce nutrients. However, analysis of the energy content of the prey and of the rates of intake of energy by redshank feeding upon the different prey revealed that the redshank would have obtained between two and three times more energy per minute by taking worms than they actually obtained by feeding almost exclusively on *Corophium*.

The lesson of this study is not merely that the energy hypothesis was disproved, but that the interpretation of results gained in one situation may appear in quite a different light when the animals are studied in another situation. In our opinion, foraging is unlikely to be based solely upon energy considerations even though it may appear to be so in certain circumstances. What is needed is not a set of models designed for the convenience of experimenters, but a basic theory, sufficiently sophisticated and sufficiently flexible to encompass the variety of species and situations found in nature. It appears to us that something akin to the economics of consumer choice may go a long way towards fulfilling this requirement.

6.1.3 Economics of animal choice

What might an economist have to say about the redshank foraging situation? First, it is obvious that the energy content of the prey is not the only factor of interest to the redshank. Let us suppose that there are two factors x_1 and x_2 which are present in different proportions in the two prey. For convenience, we may call these energy and nutrient, without necessarily implying any empirical veracity. When the redshank eat worms, we assume that they gain nutrient and energy in a fixed proportion, and when they eat shrimp (*Corophium*) this proportion is different. If we plot the consequences of eating on a graph of energy versus nutrient, we see that consumption of each prey produces a characteristic vector, as illustrated in Fig. 6.5.

Next the economist would ask about the price of the two commodities. As we have seen, the redshank have to put in much more effort to obtain the same energy from shrimps as from worms. The price of shrimps in energy terms is twice that of worms. In Fig. 6.6 this price difference is represented as a budget line, indicating the energy expenditure that is required to obtain given amounts of the two prey.

Next, the economist would expect the two factors to enter as arguments of a two-variable utility function, and Fig. 6.7 illustrates two curves obtained from the hypothetical utility function invented

Optimal decision-making 97

	Worms	Shrimps
Energy	7	3
Nutrient	3	7

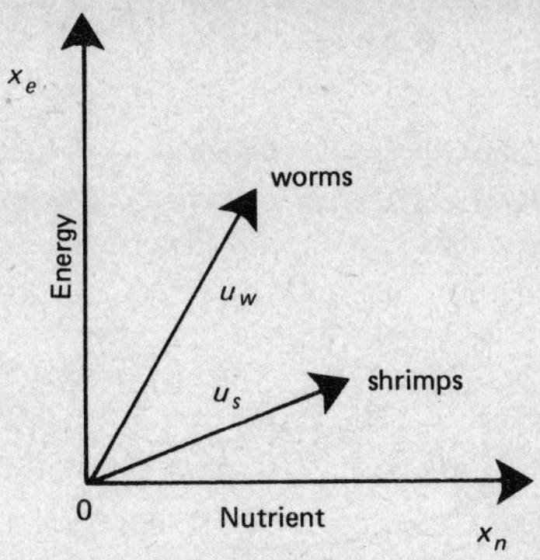

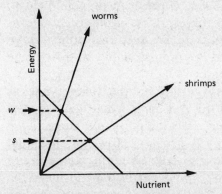

Fig. 6.5 Hypothetical composition of worms and shrimps (left), and vectors resulting from consumption of worms (u_w) and shrimps (u_s) (right).

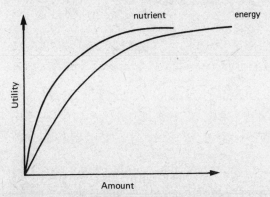

Fig. 6.6 Budget line based upon energy prices of worms and shrimps, superimposed on Fig. 6.5.

Fig. 6.7 Hypothetical utility functions for nutrient and energy.

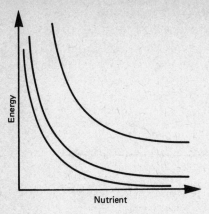

Fig. 6.8 Iso-utility functions for nutrient and energy, based upon Fig. 6.7.

for the sake of this exercise. These two curves can be combined to give roughly hyperbolic indifference curves, or iso-utility functions, as illustrated in Fig. 6.8. Every point on a given iso-utility curve represents the same total utility that the animal would gain from the combination of energy and nutrient.

If we superimpose the iso-utility curves on the budget line as shown in Fig. 6.9 we can see that the redshank would gain greater utility by taking shrimps alone than by taking worms alone, even though the energy price of shrimps is greater. However, it would appear that the redshank could do even better by taking a mixture of worms and shrimps. In practice this would partly depend upon the cost of changing between the two prey. Thus if redshank, feeding

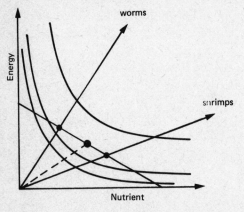

Fig. 6.9 Iso-utility functions of Fig. 6.8, superimposed on Fig. 6.6. Dotted line indicates best possible mixture of worms and shrimps, assuming that there is no cost of changing between them.

upon shrimp, had to fly a mile to obtain worms, we should expect them to change less often than if the worms were available nearby. Larkin and McFarland (1978) argue that the cost of changing should always be allocated by the animal to the cost of the activity that the animal is changing to, and they provided evidence that this is the case for doves (*Streptopelia risoria*) choosing between food and water. In the present case, this principle would mean that, for redshank feeding on shrimps, the energy cost of changing to worms would be added to the price of worms – up to the point that the redshank started feeding upon worms, when the situation would be reversed.

We can imagine that, at the outset of a foraging period, the animal has a certain amount of energy available. Changes in this quantity cause the budget line to shift in a parallel fashion, but this shift will generally lead to little change in the optimal mixture, as indicated in Fig. 6.10. Price changes, on the other hand, alter the slope of the budget line, as shown in Fig. 6.11, and this does cause a considerable shift in the optimal mixture. If the energy cost of changing is added to the price of the future activity, then the current activity will maintain a higher utility for a greater proportion of the time. In general, in calculating the costs and benefits associated with the behaviour of animals, we have to take account not only of the activities themselves but also of the cost of changing between activities (Larkin and McFarland, 1978; see also section 8.2.3).

So far, in likening the animal to an economic consumer, we have made a distinction between cost or utility as the common currency of decision-making and energy-cost or price. Although energy-cost is

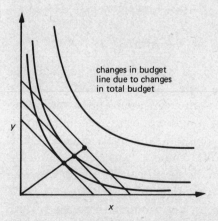

Fig. 6.10 Changes in the budget line due to the initial amount of energy available to the animal.

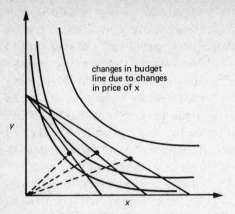

Fig. 6.11 Changes in the budget line due to changes in the price of X. Dotted lines indicate shifting optimal preferences.

common to all activities, it should not be treated as a common metric for utility, cost, or contribution to fitness. This is a distinction which is extremely important, but has not always been made clear in behavioural ecology. Pursuing the economic parallel, we can regard energy as analogous to money. The animal can earn it by foraging and may spend it upon various other activities. Over and above the basic continuous level of metabolic expenditure, the animal can save money by hoarding food or depositing fat, or can spend money upon various activities. When the price of activities is high, the animal is subject to a tight budget constraint, and when the price is reduced the animal experiences an increase in real income, and the budget constraint is relaxed (see section 7.2).

6.1.4 Elasticity and the availability of 'substitutes'

In the previous section we considered what economists call the 'income effect': an increase in the price of a commodity can be thought of as a decrease in the consumer's income under conditions of constant prices. Another result of one commodity becoming more expensive is that other commodities become relatively cheaper. This is known as the substitution effect. We would expect the demand for a commodity whose price has risen to be elastic if there were an acceptable substitute available. For example, Lea and Roper (1977) found that the elasticity of demand for food pellets increased when sucrose was also available. The limiting case of substitutability arises when there are two ways to obtain the same commodity. For example, imagine that an animal has the choice of working for food on two fixed ratio schedules, A and B. Schedule B is not changed, but schedule A starts off with a lower ratio than B, but has its ratio increased over a number of sessions. An 'economic' animal should choose schedule A as long as its ratio is smaller than that of schedule B, but should switch to schedule B when its ratio becomes relatively smaller. This behaviour was first observed by Herrnstein (1958).

When the alternative is not identical to the commodity under consideration, the elasticity will depend on how good a substitute it is. We believe that the state-space approach provides a natural framework for representing this 'goodness'. The point is that although the results derived in Chapter 7 are based on regarding activities as having consequences for just one axis of the state-space, this is a simplification for the purpose of exposition. As soon as each activity is represented as influencing several axes, it becomes possible to relate activities to each other by considering the similarity of their consequences. If two activities have many consequences in common, we would expect them to be good substitutes for each other. The exact relationship will depend on both the shared consequences and their relative magnitudes.

As an example, imagine that a predator such as the redshank discussed above can feed on two sorts of prey, W and S, both of which have consequences for two axes x_e and x_n, as is shown in Fig. 6.5. By doing activity u_W, the animal obtains mostly E, but also some N. Activity u_S gives mostly N, but also substantial amounts of E. Activity u_S is clearly a reasonable substitute for activity u_W. If, however, the angle between the activity vectors were larger, then activity u_S would not be such a good substitute for u_W. Instead of plotting E and N on the main axes, we can plot the activities u_S and u_W orthogonally. By doing this we effectively prise apart the consequence vectors illustrated in Fig. 6.5 and stretch out the iso-utility curves (Fig. 6.9). Clearly, there will be more stretching required the more the two activities are substitutes for each other. In general, the iso-utility curves plotted in relation to activities will be less convex the more the activities are substitutes for each other.

When two activities are good substitutes for each other, fluctuations in the price of one will have a marked effect upon the amount purchased, i.e., elasticity of demand will be high. When the two activities are poor substitutes for each other, as are foraging for food and for nest material, the iso-utility curves will be convex, and fluctuations in the price of one activity, say foraging for food, will have little effect upon demand. This is a well known effect in consumer economics.

This example shows that it is possible to represent the benefits of behaviour either in direct behavioural terms (i.e. in a space with behavioural axes), or in terms of the consequences of behaviour (i.e. in a space with axes representing the animal's state). Both approaches are capable of giving adequate representations of particular situations. However, there are a number of phenomena, such as realizability and observability (see section 5.4) that arise out of the fact that behaviour can have simultaneous consequences affecting more than one state variable. For example, in Fig. 6.5 the area enclosed by the activity vectors represents the realizable consequence space. If the optimal solution to the problem were to lie outside this area, then the behaviour would be constrained simply by the proportions of the constituents of the diet. These and similar problems are discussed at length in sections 2.2 and 2.3. We can content outselves with the observation that the state-space approach provides a natural way to handle such phenomena, whereas the behaviour-space approach can only cope by introducing *ad hoc* assumptions. This is also evident from considerations of the interactions between behaviour and acclimatization, which we discuss in section 9.3.

6.2 Decision-making mechanisms

There are three main stages in reaching a satisfactory understanding of decision-making. Firstly, a maximizing principle must be established; secondly, there must be a recognition that there will inevitably be some trade-off among various aspects of the problem; and thirdly, a set of optimality criteria must be formulated. Let us take, as an everyday example, a university committee set up to review applications for a lectureship. Their maximizing principle is to choose the best candidate, but opinions as to the implications of this principle may vary from one university to another. In some universities strength of condidature is seen in academic terms, but in others it may be seen in political terms, or the best candidate may be the one who is most closely related to the president of the university. Let us assume that academic ability is the principle. Normally the requirements of a lectureship are such that there is inevitably a trade-off between the teaching experience T and the research ability R of any given candidate. The committee should attempt to assess these attributes independently, and then combine these to produce an overall strength of candidature. Suppose that they are able to score R and T on a ten-point scale, and that for candidate A, $T=9$ and $R=1$; while for candidate B, $T=2$ and $R=7$. Any assessment of overall strength of candidature must rest on the optimality criterion used to evaluate teaching ability in relation to research ability. For example, if the optimality criterion was $C=T+R$, then candidate A would score $C=9+1=10$, and candidate B would score $C=2+7=9$. Alternatively, if $C=T\times R$, then A scores $C=9\times 1=9$ and B scores $C=2\times 7=14$. Thus by altering the optimality criterion, without changing the scores, it is possible to come to a different conclusion as to the best candidate.

What factors influence the optimality criteria in a university context? Factors such as government attitude, financial considerations, etc., may determine the universities' overall policy concerning teaching and research. As far as the decision-making body is concerned, these are ecological factors which have to do with the function of a university in the country as a whole.

In the study of decision-making in animals, maximizing principles are usually seen in terms of *cost* – a more-or-less sophisticated index of the action of natural selection. The trade-off is between activities that are mutually exclusive in the sense that they cannot be performed simultaneously, and the optimality criteria are embodied, either in a set of decision-rules or in some kind of objective function.

Let us suppose that the strength of candidature for feeding, which we may call the feeding tendency, results from a trade-off between internal hunger and strength of food cues. The same feeding tendency can result from different combinations of these variables. The line joining all points of equal candidature strength is called a motivational isocline (see section 1.3) as the animal has to compare its feeding tendency against tendencies for other types of behaviour, it is important that the criteria determining the feeding tendency be related to the animal's needs as a whole. An animal in which the feeding tendency is too dominant over other aspects of behaviour or is too easily swamped by competing tendencies will be at a disadvantage compared to other animals. Therefore we would expect the optimality criteria (shape of the isoclines) for feeding to be designed by natural selection in accordance with the animal's ecological circumstances. For instance, where food availability is erratic, more emphasis should be given to cue strength, while the emphasis attached to hunger should be related to the animal's physiological tolerance (McFarland, 1976).

The point we wish to stress is that, however we investigate proximate causal mechanisms of decision-making in animals, we are sooner or later going to come up against the trade-off problem. Because there is a trade-off in terms of natural selection, which can be expressed in terms of a cost function (see section 6.1.1), there must also be a trade-off that is embodied within the decision-making mechanisms of each individual animal.

6.2.1 Objective functions

Objective functions specify the notional costs and benefits of the state of the animal and its behaviour. In other words, the objective function describes the relationship between the different behavioural options in such a way that the options are evaluated in terms of their contribution to the entity that is maximized by the individual in the decision-making process.

At this point we have to distinguish between the objective function as a property of the individual animal and the objective function as seen by an investigator. We will use the term goal function to refer to properties of the individual animal, and retain the more general term objective function to include the view of an investigator.

The goal function is envisaged as a property of the individual animal and can be expected to differ from one individual to another. Goal functions might be expected to be similar in related individuals and in individuals living in similar environments. We argue later (Chapter 10), however, that goal functions cannot be modified by learning, except in a pre-programmed manner.

So far we have put forward the view that animals are machines which are designed to maximize a particular entity, one of the components of which is a goal function that provides the trade-off criteria between different behavioural options. The exact nature of the maximized entity will depend upon the ways in which the animal's behaviour is designed to alter the causal state of the animal (see Chapter 5).

The goal function will include those properties of the animal which are relevant to the notional costs associated with the various causal states and with the various aspects of behaviour. These may include physical properties of the animal as well as evaluations represented in the brain. For example, a person travelling across town on foot has to make decisions as to when to rest, to walk, to jog, to pause, etc. The goal function will include factors associated with the person's energy reserves, the risks of crossing roads, etc. It would also include some more physical factors, such as the length of the person's legs. The natural walking pace is partly determined by length of leg, because this length influences the natural frequency of the locomotory system and thus affects the walking pace at which the rate of energy expenditure is least. Optimal decisions between jogging, walking, pausing and resting will inevitably be affected by this, because there will be a trade-off between rates of energy expenditure and speed of travel.

A completely deterministic decision-maker will have a set of decision rules that operate automatically in accordance with state of the causal factors. We can imagine that the state space is divided up into regions corresponding to different activities. The state changes continuously as a result of the behaviour, and when the trajectory crosses from one region to another the behaviour changes. The decision rules specify the positions of the boundaries between the different regions. The decision rules prescribe the optimal behaviour as evaluated by the goal function. The direction of the trajectory, however, depends upon the consequences of the behaviour. These have characteristic effects upon the animal's physiology, which in turn affects the motivational state. The situation is similar to that of industrial process control, in which an industrial plant is controlled by a computer in accordance with a set of rules. The manipulations made by the computer correspond to the behaviour of the animal, the industrial plant corresponds to the animal's physiological system, the state of which is monitored by the computer. We can image an animal as controlling its physiological state through the medium of behaviour. Just as in an industrial plant, some aspects of the physiological system are self-regulating, but intervention is necessary to achieve certain objectives, which may change with changing circumstances.

It is convenient to refer to the physiological machinery relating behaviour to the consequent changes in motivational state as the system plant. The behaviour that is observed depends jointly upon the nature of the plant and the decision-rules that determine what behaviour is performed as a result of changes in the state of the plant. For the observed behaviour to be in any biological sense optimal, the decision-rules must be such as to conform with some goal function.

Optimal behaviour is not a logical necessity and indeed the evidence suggests that in some primitive animals the decision-rules are relatively rigid. For example, the lugworm *Arenicola*, which lives in burrows in sandy beaches, carries out its activities in a fixed sequence A, B, C, D, A, B, C, D, with clockwork regularity regardless of environmental changes (Wells, 1966). The mollusc *Pleurobranchaea californica* has a fixed hierarchy: egg-laying → feeding → mating → other activities (Davis, Mpitsos and Pineo, 1974). In an optimizing animal the decision rules are such that the behaviour of the animal maximizes some entity and this requires that there are some inherent optimality criteria to provide the trade-off between the inevitable alternative courses of action. These criteria are provided by the goal function.

Let us now look at the situation from an investigator's point of view. We must suppose that the investigator has been able to characterize the system plant by means of a set of plant equations, which provide a description of the animal's causal factor state (at least in its important respects) and are capable of predicting what changes in state will result from a particular behaviour pattern.

The next question the investigator must ask is whether or not the animal is an optimizing machine. If the investigator can establish that the decisions made by the animal are transitive, then it can be concluded that the animal is an optimizing machine which operates on the basis of some particular goal function. In practice it may be difficult to establish that the decisions are transitive. While in some cases it has been possible to demonstrate strong transitivity (e.g., Baerends and Kruijt, 1973), in other cases only weak stochastic transitivity has been obtained (e.g., Navarick and Fantino, 1974). Part of the problem is that the choice experiments should be repeated under identical conditions and this is never possible once an animal has made a choice. McFarland and Sibly (1975) argue that the axioms that are necessary to prove transitivity are acceptable to every biologist.

An alternative approach is to assume that the animal is an optimizing machine and to try to find its goal function. This is called the inverse-optimality approach. Basically the investigator forms an hypothesis about the goal function and tests this hypothesis by using optimality theory to predict the behaviour of the animal under a given set of circumstances. The methodology of this approach is discussed in Chapter 9 and, in order to avoid confusion, the hypothesis goal function is called an objective function. The aim of the inverse-optimality exercise is to establish that a particular objective function can account for the observed behaviour of the animal.

The question may then arise as to whether there is another

objective function which could also account for the behaviour. This is called the objective-uniqueness problem.

6.2.2 Optimal behaviour

We can now discuss what it means to say that an animal is behaving optimally. There are a number of different meanings that may be attached to the notion of optimal behaviour. Firstly, as mentioned in section 6.2.1 an animal may or may not be an optimizing machine in the sense that its behaviour conforms with some goal function. We may postulate that an animal is behaving optimally in the sense that its behaviour maximizes some entity as embodied in its goal function. A further distinction, which will be discussed later (section 9.2), is that the animal may be behaving in an overall optimal manner or may only be behaving optimally with respect to a given set of constraints.

Secondly, an animal may appear to be behaving optimally if its observed behaviour can be shown to conform to a particular objective function. Once again, there may be a distinction between the overall optimal behaviour and the constrained optimum. The question now arises as to whether an animal that can be shown to be behaving optimally with respect to an objective function is necessarily behaving optimally with respect to a goal function. This question will be discussed in section 9.2.

Thirdly, an animal may be supposed to be behaving optimally with respect to natural selection. As mentioned in section 6.1.1, this is tantamount to asking whether the behaviour conforms to a particular cost function. If the animal were perfectly adapted to its environment its goal function would be identical with the cost function characteristic of that environment. This will rarely be the case, for reasons outlined above (section 6.1.1). Normally, the goal function will approximate the cost function, since if it were to deviate too much the animal would be distinctly maladapted. We can think of the goal function as representing the notional costs and benefits related to behaviour. Normally, the animal will have no way of knowing by how much these deviate from the real costs and benefits, because inclusive fitness, the ultimate index of cost, is remote from the immediate consequences of behaviour. In some cases, however, the animal may be able to obtain some feedback about the fitness of its own behaviour. For example, a bird may have a tendency to build its nest close to the ground where the microclimate is favourable. In the ancestral environment there may have been few predators. If the nest is repeatedly disrupted by predators, the bird may rebuild in a safer place. There are many species that attempt second broods (Lack, 1966). There are some which shift the nest site following the loss of the clutch. The question arises whether the choice of a new nest site involves criteria different from those of the first nest.

6.3 Summary

Analogies between economics and ethology point to definitions of a number of parallel concepts. Thus *cost* is a notional measure of the contribution to fitness of the behaviour of an animal, and of its motivational state, whereas *utility* is a notional measure of the psychological value of goods, leisure, etc.

While the behaviour of an animal maximizes some *objective function*, economic man maximizes a *utility function*. To attain maximum fitness in a given environment an animal's behaviour should maximize the *cost function* characteristic of that environment. To attain maximum social value human behaviour should maximize a *policy function*.

In likening an animal to an economic consumer, a distinction should be made between the common currency of decision-making, which is utility or cost, and the price or energy cost of behaviour. Energy can be regarded as analogous to money, the availability of which determines the budget constraints. *Demand functions*, in both human and animal economics, relate the selection of behaviour to its price. When different activities are good substitutes for each other demand tends to be elastic, and *vice versa*. Examples of demand functions in animal behaviour are discussed.

In investigating decision-making in animals it is important to distinguish between the objective function as a property of the individual animal and the objective function as seen by an investigator. To avoid confusion the term *goal function* can be used to denote the attributes of the individual animal, and the term objective function reserved for expressions of the investigator's opinion. Both these concepts are mathematically analogous to the cost function, but when the distinctions are made in biological terms it becomes clear that a variety of meaning can be attached to questions about optimal behaviour.

7 Static optimization

As described in Chapter 6, the concept of optimality is now widely used in behavioural studies. In this chapter we outline the nature of an optimization problem. We take as our starting point the idea that an animal can use behavioural means to regulate its physiological state. Under favourable conditions an animal may be able to do all that it would 'like' to do, but when times are hard it may have to make a decision about what to leave undone. In order to predict how an animal should allocate its time under such circumstances, we need to know the costs associated with various deviations from the 'ideal' state. This amounts to defining the optimality criterion for this problem. To complete the characterization of the problem we also need the set of possible solutions. This set depends upon the constraints that are imposed. For example, in the case that we consider below, the animal can be constrained by a reduction in the total time available for performing behaviour. We elaborate these points in describing the analysis presented in section 7.1. Our assumption is that an animal allocates its time in such a way that the resulting cost at the end of the day is minimized. Section 7.2 presents the optimal solutions from which we develop a concept of elasticity analogous to that used in economics. This analogy is explored in section 7.3. We conclude the chapter with a section on the derivation of the matching law from an optimality principle.

7.1 The problem

7.1.1 Movements in state space

In Chapters 1 and 2 it was shown how an animal's state could be represented in an n-dimensional space. The state can be thought of as a specification of the value of n variables, where n is large enough to characterize the animal for our purposes. In this chapter we consider a simple model describing how an animal should behave in order to control its state in an optimal way. The particular arguments we use are based on avoiding lethal regions of state space, and hence apply only to physiological variables. It seems reasonable to assume, however, that a similar sort of approach can be used for the variables which do not have lethal boundaries.

The model incorporates a very simple relationship between behaviour and state. It is assumed that when the animal is performing activity u_i, the rate of change of state is given by

$$\dot{x}_i = -r_i, \tag{7.1}$$

i.e., activity u_i, has consequences for only axis x_i. As we mentioned in section 6.1.3, there are good reasons for preferring a vector formulation, but the scalar model is easier to work with and permits a clearer development of the optimality argument. The value of r_i in this model represents the 'returns' the animal gets from activity u_i.

7.1.2 The optimality criterion

It is generally possible to pose an optimality criterion as a maximization or a minimization. This is because maximizing some function F is equivalent to minimizing $-F$. In this section we argue that there is a cost associated with being in each particular physiological state. Ultimately this cost must be related to Darwinian fitness (see Chapter 6), but in this chapter we concentrate on survival rather than on reproduction. Despite this limitation we refer to the optimality criterion as the cost function.

It seems reasonable to assume that risk of death must increase steeply the nearer a variable is to its lethal boundary. For example, it is obviously dangerous to allow hunger to approach lethal levels if a future food supply is not guaranteed. Some support for this view is provided by the following model, based on the 'gamblers's ruin' problem (see, for example, Feller (1957), p. 317).

Consider a physiological variable $\tilde{z}$ which takes values between two lethal boundaries $\tilde{z}_0 = 0$ and $\tilde{z}_1 = L$. Assume that when the animal is not performing behaviour relevant to this variable, the variable moves either to the right or to the left with the probability of a half. (It is simplest to take time to be discrete and to imagine a move occurring at each instance of time.) If M is the expected number of moves required for $\tilde{z}$ to reach $\tilde{z}_0$ or $\tilde{z}_1$, then M is given by the following formula (Feller, 1957):

$$M = \tilde{z}(L - \tilde{z}) \tag{7.2}$$

By the symmetry of this model, M must be at a maximum when $\tilde{z}$ is equidistant from $\tilde{z}_0$ and $\tilde{z}_1$, i.e., when $\tilde{z} = L/2$. This suggests introducing a new variable $z = \tilde{z} - L/2$, i.e., z is the distance (displacement) from the optimal state. Equation (7.2) then becomes

$$M = (z + L/2)(L - z - L/2)$$
$$= L^2/4 - z^2$$

M gives some indication of the safety of being at z. By the claim made above, maximizing M is equivalent to minimizing $-M$, which should be some indication of risk. As $L^2/4$ is a constant, this suggests a cost function of the form

$$c(z) \propto z^2$$

A similar sort of argument can be put forward for the case in which changes in state confirm to a normal distribution, as is shown in Fig. 7.1. The probability that the state crosses the lethal boundary is given by area A in the figure. When the state is near to the boundary this area is approximately proportional to the square of the distance from the boundary. (See Appendix 7.1 for a justification of this claim.)

Although these arguments encourage the view that the cost is proportional to x^2 (i.e. the cost function is quadratic), they by no means constitute a proof of its validity. The best justification would seem to be that any reasonable function must be convex, and the quadratic is a convenient approximation.

When more than one state is being considered, some assessment of the total cost $C(x)$ must be made. If $C(x)$ can be represented as the sum of the cost associated with each x_i in x, then $C(x)$ is said to be

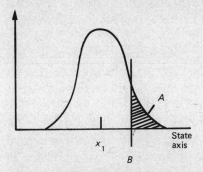

Fig. 7.1 The probability P of the state x_1 moving to a new position along the state axis. Area A is the probability of crossing boundary B. (After Houston and McFarland, 1980.)

separable. Sibly and McFarland (1976) show that separability is approximately equivalent to probabilistic independence of the various factors. Independence in this context means that the risk associated with the value of one variable is independent of the values of the other variables. This would be the case, for example, if the probability of death from heat stress were unaltered by the nutritional condition of the animal. This is not to say that nutritional condition does not influence the probability of survival but that the effect of body temperature on survival is independent of nutritional condition.

It is assumed that separability holds for the model described in this chapter so that cost $C(x)$ of being in state x is a weighted sum of the squares of the displacements that constitute x. For example, if $x = x_1, x_2, x_3$ then

$$C(x) = \frac{x_1^2}{Q_1} + \frac{x_2^2}{Q_2} + \frac{x_3^2}{Q_3} \tag{7.3}$$

The weighting parameter Q_i will be referred to as the resilience of the variable x_i. (See Houston and McFarland, 1980.) The optimality criterion then amounts to requiring the animal to spend its time in such a way that the displacements at the end of the day result in the smallest possible cost, which will be zero under favourable circumstances. There are two reasons for taking the period of time to be a day. Firstly, it is an obvious choice on biological grounds, in that circadian patterns are so marked. Secondly, the quadratic cost function is based on risks when it is assumed that behaviour is not in control of the state. The obvious interpretation of this is that the animal has to reach a state at the end of the day which minimizes its risks during the night.

To complete the specification of the optimization problem we use equation (7.1) to link the animal's behaviour to consequences for its state. If during a day the duration of time spent performing activity u_i is d_i, then the total consequence for axis x_i is $d_i r_i$. In other words, if x_i began the day at a value $x_i(0)$, its value at the end of the day, $x_i(T)$, would be given by $x_i(T) = x_i(0) - d_i r_i$. (As we will always be concerned in this chapter with the state at the end of the day we will continue to write x_i for $x_i(T)$.) It is possible to remove all explicit mention of $x_i(0)$ by imagining an 'ideal' time budget which results in no deficit on any axis at the end of the day. Representing the parameters d_i and r_i of this budget by D_i and R_i respectively, we can now measure displacements from $D_i R_i$ instead of from $x_i = 0$. To summarize the argument:

$$x_i(0) - d_i r_i = x_i$$
$$x_i(0) - D_i R_i = 0$$
$$\therefore \quad x_i = D_i R_i - d_i r_i \tag{7.4}$$

All that remains to be done is to establish the constraints. If the total duration of the day is D, then as the animal must always be doing something and cannot do two things at once (by definition of activity; see section 1.2), the basic constraint must be

$$d_1 + d_2 + \ldots + d_n = D$$

(We will generally assume that equality holds in expressions of this form. This amounts to saying that the optimum solution uses all the time available. This is certainly realistic for the particular cases we consider, and it also simplifies the theoretical issues.)

We now find the optimal choice of d_i when the animal is forced away from the ideal budget $D_i R_i$. This could be brought about in two ways:

(1) Time constraint. The basic constraint is made more severe by reducing the daily time available from D to $D - T$.

(2) 'Returns' constraint. An activity becomes less profitable, i.e. a returns parameter R_i is reduced to r_i.

7.2 The optimal solution

After some general remarks on optimization, we present a detailed derivation of the optimal solutions for the two cases outlined above. We then show how both results can be derived much more easily by the introduction of a Lagrange multiplier.

7.2.1 Unconstrained optimization

An intuitive idea of the analytic approach to optimization can be obtained from the example illustrated in Fig. 7.2. This figure shows a possible relationship between cost C and a variable x. At the point b, which minimizes the cost, it can be seen that dC/dx (i.e., the slope of the tangent to the curve) is zero. Such points are said to be stationary. The figure shows that being stationary does not suffice to

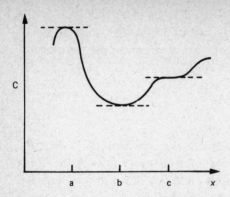

Fig. 7.2 An hypothetical graph of cost as a function of state. *a* is the maximum, *b* the minimum, and *c* a point of inflection.

characterize a minimum. The maximum *a* and point of inflection *c* are also stationary. Ignoring any problems that may arise if the minimum occurs at the edge of the set of possible solutions, we can conclude that $dC/dx = 0$ is a necessary, but not a sufficient condition for a minimum.

For cases like the one shown in Fig. 7.2, it is straightforward to find a condition that separates the minimum from other stationary points. If the second derivative (i.e., d^2C/dx^2) is calculated, it will be positive at a minimum, negative at a maximum, and zero at a point of inflection. We now apply this approach to the problem of time allocation.

7.2.2 The solutions

(a) *The constraint is the time available*
Let the time available per day be reduced by T hours. We confine ourselves to two activities, u_1 and u_2, and assume that the animal responds to the constraints by reducing the time spent doing u_1 or u_2 or both. (This assumption is relaxed in the next section.)

>Let the reduction in time spent on u_1 be called t_1
>Let the reduction in time spent on u_2 be called t_2;
>then $t_1 + t_2 = T$ and $t_1 = T - t_2$ \hfill (7.5)

Furthermore, $d_i = D_i - t_i$ and hence, from equation (7.4)

$$x_i = R_i(D_i - (D_i - t_i))$$
$$\therefore \quad x_i = R_i t_i \hfill (7.6)$$

From our previous arguments, we assume that the cost function is quadratic and separable (see equation (7.3)) and that t_1 and t_2 are to be chosen such that C is minimized. From equation (7.6)

$$x_1 = R_1 t_1 \quad \text{and} \quad x_2 = R_2 t_2 \tag{7.7}$$

t_1 can be eliminated from equation (7.7) by the relationship given in equation (7.5)

$$x_1 = R_1(T - t_2) \qquad x_2 = R_2 t_2 \tag{7.8}$$

From equation (7.8) and the definition of C,

$$C = \frac{(R_1 T - R_1 t_2)^2}{Q_1} + \frac{R_2^2 t_2^2}{Q_2}$$

$$= \frac{R_1^2 T^2 - 2 R_1^2 T t_2 + R_1^2 t_2^2}{Q_1} + \frac{R_2^2 t_2^2}{Q_2}$$

To find the value of t_2 which minimizes C we differentiate with respect to t_2 and equate to zero:

$$\frac{dC}{dt_2} = \frac{-2 R_1^2 T + 2 R_1^2 t_2}{Q_1} + \frac{2 R_2^2 t_2}{Q_2} = 0$$

(note that this is a minimum: $d^2 C/dt_2^2$ is positive)

$$\therefore \quad \frac{t_2(Q_2 R_1^2 + Q_1 R_2^2)}{Q_1 Q_2} = \frac{R_1^2 T}{Q_1}$$

$$\therefore \quad t_2 = \frac{R_1^2 T Q_2}{Q_2 R_1^2 + Q_1 R_2^2} \tag{7.9}$$

A graphical approach to this problem is given in Fig. 7.3.

(b) *The constraint is on the consequences of an activity*
Once again, only two activities (u_1 and u_2) will be considered. In this case, instead of reducing the time available for behaviour, the behaviour is made less effective: the animal achieves less for the same amount of time spent. This can be represented in the following way.

The animal usually spends a time D_i on activity u_i, and the consequences are given by $D_i R_i$. R_i is now reduced to some value r_i. If D_i is increased to compensate, then less time will be available for the other activity, because there is a limited amount of time, D (=1 day) at the animal's disposal. Table 6.1 summarizes the situation. Now that we have obtained an expression for the deficit, we

Table 7.1 Effects on behaviour of altering consequences

	u_1	u_2	
Usual amount of time spent on u_i	D_1	D_2	$D_1 + D_2 = D$
Amount of time spent after variation of consequences	d_1	d_2	$d_1 + d_2 = D$
Consequences of D_i before variation	$D_1 R_1$	$D_2 R_2$	
Amount of time given up, t_i	$t_1 = D_1 - d_1$	$t_2 = D_2 - d_2 = d_1 - D_1$	
Deficit	$D_1 R_1 - d_1 r_1$	$R_2(d_1 - D_1)$	

proceed as in case (a). The cost function (see equation (7.3)) is now:

$$C = \frac{(D_1 R_1 - d_1 r_1)^2}{Q_1} + \frac{(R_2 d_1 - R_2 D_1)^2}{Q_2}$$

$$= \frac{D_1^2 R_1^2 + d_1^2 r_1^2 - 2 D_1 d R_1 r_1}{Q_1} + \frac{R_2^2 d^2 + R_2^2 D_1^2 - 2 R_2^2 d D_1}{Q_2}$$

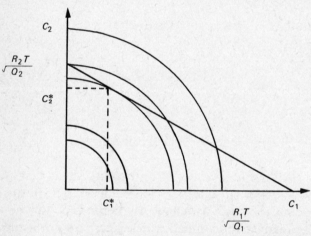

Fig. 7.3 A two-dimensional space showing isoclines of equal cost. These are circles given by $(R_1/\sqrt{Q_1}) + (R_2/\sqrt{Q_2}) = $ constant. The optimization problem arising from reducing the available time by T units can be approached graphically as follows: as $T = t_1 + t_2$, the possible displacements must be on a straight line joining $x_1 = R_1 T$ (i.e. $t_2 = 0$, $t_1 = T$) and $x_2 = R_2 T$ ($t_1 = 0$, $t_2 = T$). The corresponding line in this space will be from $c_1 = (R_1 T \sqrt{Q_1})$ to $c_2 = (R_2 T \sqrt{Q_2})$. The values c_1^* and c_2^* that minimize the cost are obtained by finding the isocline of lowest cost which is compatible with the constraint. This is indicated by the dotted lines. (After Houston and McFarland, 1980.)

To find the value of d which minimizes C, we differentiate C with respect to d and equate to zero:

$$\frac{dC}{dd} = \frac{2dr_1^2}{Q_1} - \frac{2D_1R_1r_1}{Q_1} + \frac{2R_2^2d}{Q_2} - \frac{2R_2^2D_1}{Q_2} = 0$$

$$\therefore \quad \frac{d(r_1^2Q_2 + R_2^2Q_1)}{Q_1Q_2} = \frac{D_1R_1r_1Q_2 + R_2^2D_1Q_1}{Q_1Q_2}$$

$$\therefore \quad d = \frac{D_1(Q_2R_1r_1 + Q_1R_2^2)}{Q_2r_1^2 + Q_1R_2^2} \tag{7.10}$$

7.2.3 Lagrange multipliers

The methods applied in the previous section involved using the constraint equation to eliminate a variable from the expression for cost. This procedure becomes cumbersome when the problem contains many variables and constraints. An alternative approach, based on Lagrange multipliers, is often more convenient. The multipliers can be introduced by way of a graphical representation of the optimization problem. (See for example Dixit (1976).) Figure 7.3 illustrates the time-constraint problem discussed on pages 114–15. The constraint is represented by the straight line and the cost isoclines by the curves, from which it follows that the optimal solution is (c_1^*, c_2^*). This point is characterized by the fact that the cost isocline is tangent to the constraint curve, and so their slope must be equal. To make the argument more general, consider the cost to be given by some function $C(x_1, x_2)$ and the constraint to be some function $N(x_1, x_2) = 0$. Denoting the optimum solution by (x_1^*, x_2^*), the common slope is given by

$$\frac{dC}{dx_1^*}\frac{dx_2^*}{dC} \quad \text{and} \quad \frac{dN}{dx_1^*}\frac{dx_2^*}{dN},$$

which means that

$$\frac{dC}{dx_1^*}\frac{dx_1^*}{dN} = \frac{dC}{dx_2^*}\frac{dx_2^*}{dN} \tag{7.11}$$

Let λ be value of both sides of equation (7.11), i.e.

$$\frac{dC}{dx_i^*}\frac{dx_i^*}{dN} = \lambda \tag{7.12}$$

$$\therefore \quad \frac{dC}{dx_i^*} - \frac{\lambda \, dN}{dx_i^*} = 0 \tag{7.13}$$

We can now define a new function $L(x_1, x_2) = C(x_1, x_2) - \lambda N(x_1, x_2)$ such that at the optimal solution $\dfrac{dL}{dx_i}$ is zero. The implication is that the problem of minimizing C subject to the constraint has been replaced by the unconstrained minimization of L. The optimal solutions of the previous section can now be derived as follows. In general, C is given by the following equation:

$$C = \frac{(D_1 R_1 - d_1 r_1)^2}{Q_1} + \frac{(D_2 R_2 - d_2 r_2)^2}{Q_2}$$

and the constraint is of the form

$$d_1 + d_2 = D - T.$$

Define L as

$$L = C + \lambda(D - T - d_1 - d_2);$$

then

$$\frac{dL}{dd_1} = \frac{2}{Q_1}(d_1 r_1^2 - D_1 R_1 r_1) - \lambda$$

and

$$\frac{dL}{dd_2} = \frac{2}{Q_2}(d_2 r_2^2 - D_2 R_2 r_2) - \lambda$$

Setting $\dfrac{dL}{dd_1} = \dfrac{dL}{dd_2} = 0$ and using the constraint equation gives

$$d_1 = \frac{D_1 R_1 r_1 Q_2 - D_2 R_2 r_2 Q_1 + (D-T) r_2^2 Q_1}{r_1^2 Q_2 + r_2^2 Q_1} \qquad (7.14)$$

The two solutions are now readily obtained:
(1) Time constraint. Set $D = D_1 + D_2$, $r_1 = R_1$, $r_2 = R_2$. Then

$$d_1 = \frac{D_1 R_1^2 Q_2 + D_1 R_2^2 Q_1 - T R_2^2 Q_1}{Q_2 R_1^2 + Q_1 R_2^2},$$

but

$$t_2 = d_1 - D_1 + T.$$

$$\therefore \quad t_2 = \frac{R_1^2 T Q_2}{Q_2 R_1^2 + Q_1 R_2^2}$$

which is equation (7.9).

(2) 'Returns' constraint. Set $T=0$, $D=D_1+D_2$, $r_2=R_2$

$$d_1 = \frac{D_1(R_1r_1Q_2+R_2^2Q_1)}{r_1^2Q_2+R_2^2Q_1}$$

which is equation (7.10).

The problem of n activities can be solved by considering

$$L = C + \lambda(D - T - d_1 - d_2 - \ldots - d_n).$$

For example, the 'returns' case in which $n=3$ has an optimum solution given by

$$d_1 = \frac{D_1(R_2^2R_3^2Q_1 + R_1r_1R_3^2Q_2 + R_1r_1R_2^2Q_3)}{r_1^2R_2^2Q_3 + R_2^2R_3^2Q_1 + r_1^2R_3^2Q_2}$$

7.3 Economic parallels

7.3.1 Resiliency and elasticity of demand

As explained in Chapter 6, demand curves are used in economics to express the relationship between the price and the consumption of a commodity. If the demand curve is horizontal, as shown in Fig. 6.3, the amount consumed does not vary with the price and the demand is said to be inelastic. The elasticity, η, can be roughly defined as

$$\eta = -\frac{\text{percent change in amount consumed}}{\text{percent change in price}}$$

A more precise definition is

$$\eta = -\frac{da}{dp} \cdot \frac{p}{a}$$

where p is the price and a is the amount consumed. The range of η is from zero to infinity: if η is less than one, the demand is called inelastic, while if η is greater than one demand is said to be elastic.†

In economic theory, the consumer is constrained by having a fixed income to spend on various commodities. This suggests an analogy with the problem of section 7.1 with time taking the place of money. This idea has been explored in detail by Lea (1978), who concludes that there are important similarities between reinforcement theory

---

† Sometimes elasticity is defined as $\frac{da}{dp}\frac{p}{a}$, in which case its range is from zero to minus infinity.

and demand theory. In this context, equation (7.10) is of interest in that it can be thought of as a model of fixed-ratio experiments. An animal working on a fixed ratio of size n must make n responses to obtain a reward, so increasing n corresponds to decreasing R. Although it goes beyond the original scope of our model, we believe this analogy is worth pursuing.

To derive an expression of elasticity from equation (7.10) the behavioural analogies of 'amount consumed' and 'price' must be identified. We will identify the amount consumed, a, with the total consequences obtained by the animal (i.e. time spent on an activity multiplied by its 'return' parameter). The grounds for this procedure can be made obvious by considering feeding. If the time spent feeding is d and the food availability is r, then the amount consumed is dr. Applying this argument to equation (7.10) and dropping subscripts where this is possible without confusion yields

$$a = dr = \frac{D_1(Q_2 R_1 r^2 + Q_1 R_2^2 r)}{Q_2 r^2 + Q_1 R_2^2} \tag{7.15}$$

For a given activity, a reduction of r means that the animal obtains less from a given time spent performing the activity. This means that the price of the consequences of the activity has been increased. We represent this relationship by identifying the price p with $1/r$. Equation (7.15) can now be rewritten as

$$a = \frac{D_1(Q_2 R_1 + Q_1 R_2^2 p)}{Q_2 + Q_1 R_2 p^2}$$

$$\therefore \quad \frac{da}{dp} = \frac{D_1(K + p^2) - 2p(D_1 K R_1 + Dp)}{(K + p^2)^2} \tag{7.16}$$

where K has been written for $Q_2/Q_1 R_2^2$

$$\therefore \quad \eta = -\frac{p}{a}\frac{da}{dp} = \frac{p^3 + 2KR_1 p^2 - Kp}{p^3 + KR_1 p^2 + Kp + K^2 R_1} \tag{7.17}$$

If we assume that everything is scaled so that $R_2 = 1$, K is the relative magnitude of the resiliences. When K is very big, the term in K^2 on the bottom line will dominate and η will be small. In other words, if Q_2 is much bigger than Q_1, demand is inelastic. This means that the animal is reluctant to let its daily 'returns' drop much below $D_1 R_1$. Another way to look at this case is to find an approximation for equation (7.10) when Q_1 and R_2 are small. Under such circumstances

$Q_1 R_2^2$ can be neglected and equation (7.10) becomes

$$d \simeq \frac{D_1 Q_2 R_1 r}{Q_2 r^2} \simeq \frac{D_1 R_1}{r}$$

$\therefore \quad dr \simeq D_1 R_1$, which is a constant, i.e. demand is completely inelastic.

Equation (7.17) can be used to obtain a condition for elasticity. By definition, demand is elastic if $\eta > 1$, i.e., if

$$p^3 + K R_1 p^2 + K p + K^2 R_1 < p^3 + 2 K R_1 p^2 - K p$$

which implies

$$K < \frac{R_1 p^2 - 2p}{R_1}$$

In other words, if demand is to be elastic, K must be less than $p^2 - \frac{2p}{R_1}$.

Although demand curves showing the relationship between price and consumption are an obvious representation for an economist to use, psychologists and ethologists are likely to be just as interested in behaviour as in consumption. To see how the time spent performing an activity varies with its 'price', we must return to equation (7.10) and replace r by $1/p$:

$$d = \frac{D_1 (Q_2 R_1 p + Q_1 R_2^2 p^2)}{Q_2 + Q_1 R_2^2 p^2} \tag{7.18}$$

Equation (7.18) shows that the result of increasing p will depend on the relative magnitudes of the resiliences. This can be emphasized by replacing $Q^2/Q_1 R_2^2$ by K and differentiating with respect to p:

$$\frac{\partial d}{\partial p} = \frac{D_1 (K^2 R_1 + 2 K p - K R_1 p^2)}{(K + p^2)^2} \tag{7.19}$$

Equation (7.19) shows that the time spent on an activity will only increase with increasing price if $K^2 R_1 + 2 K p$ is greater than $K R_1 p^2$. This will generally be the case if K is big, so that $K^2 R_1$ dominates. this is, of course, just another way of deriving the condition for inelastic demand. As p becomes very big, the negative term in p^2 will reduce d. This decrease may not occur within the range of p investigated unless K is small. A case in which d rises and then falls is illustrated in Fig. 7.4.

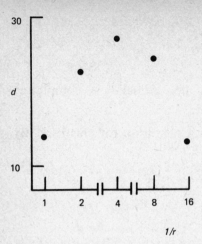

Fig. 7.4 A graph of d as a function of $1/r$ from equation 7.10, using the following values: $R = 5$, $D_1 = 10$, $k = 2/3$. (After Houston and McFarland, 1980.)

The behaviour of equation (7.18) can now be compared with results from various experiments using fixed-ratio reward schedules. The basis of the comparison is the identification of the number of responses needed to obtain a reward with the price, p. It then follows that d represents the number of responses the animal makes. Many examples of the behaviour of animals on fixed ratio schedules can be found in the review by Hogan and Roper (1978). They conclude that inelastic demand ('compensation' in their terminology) characterizes the relationship between rewards and the response ratio (the price) when food is the reward. As the ratio is increased, the number of responses also rises to maintain a roughly constant number of rewards (Hogan, Kleist and Hutchings, 1970; Collier, Hirsch and Hamlin, 1972). Other reinforcers may produce elastic demand curves. For example, Hogan *et al.* (1970) found that as the number of responses a male Siamese fighting fish (*Betta splendens*) had to make to obtain the chance to display to a mirror was increased, the fish made a roughly constant number of responses, and hence obtained fewer reinforcements, as is shown in Fig. 7.5. Elastic demand can also be found in behaviour that has an obvious regulatory function. Carlisle (1969) found that shaved rats have an elastic demand for heat reinforcements; indeed the response rates were so low that some of the rats died from cold during the experiment.

It is also possible to find examples of '∩' shaped functions. Fig. 7.6 shows data from Teitelbaum (1961) and from Mrosovsky (1968). Teitelbaum worked with rats which had been given lesions in the

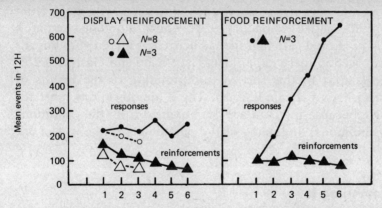

Fig. 7.5 Mean response rates of groups of *Betta splendens* reinforced for swimming through a tunnel for food reward or for opportunity to display to their mirror image. The fixed-ratio schedules denote the number of responses required to obtain a reward. O.L. is the operant (baseline) level. (After Hogan *et al.* 1970.)

ventro-medical hypothalamus, Mrosovsky with hibernating rodents. In both cases the animals overeat if food is available but will not work to obtain food.

All these results can be accounted for by equation (7.18), if we take u_2 to be some lumped representation of alternative activities with resilience Q_2. The elasticity of feeding then suggests that the resilience associated with hunger is bigger than Q_2, whereas the

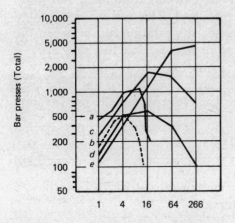

Fig. 7.6 Tests of hunger motivation in two hibernating species compared with the performance of normal and brain-damaged rats: (a) dormice, (b) ground squirrels, (c) dynamic hyperphagic rats, (d) obese hyperphagic rats, (e) normal rats. (After Mrosovsky, 1968.)

opposite could be the case for display and heat. The data from Mrosovsky (1968) and Teitelbaum (1961), which should be compared with Fig. 7.4, might indicate that the resilience associated with hunger has been increased in these experiments (see King and Gaston, 1976, for some objections to this sort of interpretation of the results of ventro-medical hypothalamic lesions). These identifications must remain tentative because the data are not obtained from the conditions assumed by the model and many of the parameters in the model are 'free'.

7.4 The matching law

In the previous section we considered the behaviour of animals on ratio schedules of reinforcement as a possible example of a static optimization problem. Results similar to ours have been obtained independently by Staddon (1979) and Rachlin (1978), who have also extended the analysis to include variable interval (abbreviated as V.I.) schedules. In Chapter 4 we define a V.I.–T sec. to be a series of fixed interval schedules with mean interval T sec. The distribution of intervals usually approximates a negative exponential, in which case the V.I. can be thought of as being generated by a constant probability of a reward being set up at each instant (see Staddon, 1977). As a consequence, the longer the interval between responses, the more likely it is that a response will be rewarded. This can be represented as a 'feedback function' that expresses the relationship between the rate u at which the animal makes responses and the rate $r(u)$ at which it obtains rewards. From what has been said about V.I. schedules, the function must start by increasing with u but must level off at the maximum possible rate. Two forms of feedback function have been considered:

Power function (Rachlin)

$$r(u) = ku^n, \qquad (7.20)$$

where $0 < n < 1$, and k is the maximum rate at which rewards can be obtained (=reciprocal of interval).

Hyperbola (Staddon)

$$r(u) = ku/(k+u) \qquad (7.21)$$

When an animal has a choice between two concurrent V.I. schedules, it can choose the response rates u_1 and u_2 to each schedule, subject to some constraint such as $u_1 + u_2 = T$.

If the total reward rate R_1 is given by $r(u_1)+r(u_2)$, then it is a simple matter to find the values of u_1 and u_2 that maximize R. We illustrate the procedure when the feedback functions are power functions:

$$r(u_1) = k_1 u_1^n$$
$$r(u_2) = k_2 u_2^m$$
$$R = r(u_1) + r(u_2) \qquad (7.22)$$

Define $L = R + \lambda(T - u_1 - u_2)$; then

$$\frac{\partial L}{\partial u_1} = nk_1 u_1^{n-1} - \lambda = 0,$$

and

$$\frac{\partial L}{\partial u_2} = mk_2 u_2^{m-1} - \lambda = 0.$$

$$\therefore \quad nk_1 u_1^{n-1} = mk_2 u_2^{m-1}$$

but

$$k_1 u_1^{n-1} = \frac{r(u_1)}{u_1}$$

and

$$k_2 u_2^{m-1} = \frac{r(u_2)}{u_2}.$$

$$\therefore \quad \frac{nr(u_1)}{u_1} = \frac{mr(u_2)}{u_2}$$

$$\therefore \quad \frac{u_1}{u_2} = \frac{nr(u_1)}{mr(u_2)} \qquad (7.23)$$

When $n = m$, equation (7.23) reduces to the matching law, which has been shown to have considerable generality (de Villiers, 1977).

Both Rachlin (1978) and Staddon and Motheral (1978) interpret this result as showing that matching is the optimal strategy, but this view is challenged by Heyman (1979) who points out that the animal can get a reward on one schedule by working on that schedule or by returning to the schedule after spending some time working on the other schedule. By assuming additivity in equation (7.22), Rachlin (1978) and Staddon and Motheral (1978) ignore this problem (but see Staddon and Motheral, 1979). Heyman and Luce (1979a) have argued that maximizing reward rate does not result in matching, and this has produced a controversy (see Rachlin 1979; Heyman and Luce 1979b). Houston and McNamara (1981) show that matching does not hold at the optimum, and does not suffice to specify the animal's behaviour uniquely.

7.5 Summary

Formulation of optimality criteria can be based upon considerations of the stability and viability of the animal's (causal factor) state. We introduce the notion of a quadratic cost function based on arguments concerning the probability that the state will cross a lethal boundary in the state space.

Once an optimality criterion has been formulated it is possible to propose optimal time budgets for situations involving constraints such as a limit on the total time available, or a limit on the rates at which particular activities can be performed. The results of such calculations show that the problem is closely related to the economics of demand, and we discuss some interesting analogues of demand functions in animals.

Appendix 7.1

An approximation for the probability of crossing a lethal boundary.

Although the normal probability distribution is defined between $+\infty$ and $-\infty$, we will develop an approximation based on considering the tail of the distribution beyond the boundary to be a triangle (see Fig. 7.A). In other words, the probability of a displacement greater than k (see figure) is assumed to be zero. A reasonable approximation can be obtained by taking k to be 3 times the standard deviation.

From the figure, the probability of crossing the boundary is given by the area A. By elementary geometry,

$$A = \tfrac{1}{2}bh$$

and

$$h = b \tan \theta$$

$$\therefore \quad A = \tfrac{1}{2}b^2 \tan \theta$$

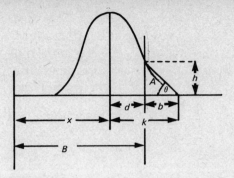

Fig. 7A.

but

$$b = k - d$$

$$\therefore A = (k^2 + d^2 - 2kd)\frac{\tan \theta}{2}$$

Now for any distribution, θ is constant, and therefore so is $\frac{\tan \theta}{2}$. As a result, we can write

$$A \propto (k^2 + d^2 - 2kd)$$

which shows that A depends on terms in d^2 and d.

From the diagram, d is given by $B - x$.

$\therefore \quad A \propto x^2 + 2x(k - B)$ where x is the distance of the state from the origin.

This result is only an approximation and depends on the range over which d is varied, but it can be quite accurate, as the following table shows:

d	A (from tables)	A (from approximation)
2·0	0·0228	0·0228
2·1	0·0179	0·0185
2·2	0·0139	0·0146
2·3	0·0107	0·0112
2·4	0·0082	0·0082
2·5	0·0062	0·0057

The approximation is given by

$$A = 0 \cdot 0228(9 + d^2 - 6d)$$

8 Dynamic optimization

The previous chapter presented some simple models using static optimization, in which time does not enter directly into the problem. In this chapter we consider dynamic problems, in which the action taken at any time has consequences which are evaluated over some period in the future. In fact dynamic problems can be defined more generally as a search for '... a set of design parameters, which are all continuous functions of some other parameter, that minimizes an objective function subject to the prescribed constraints', (Rao, 1978). In other words, the problem involves looking at the cost associated with various paths through some parameter space. The optimum solution is the path along which the total cost is least. Finding this total cost involves the mathematical operation of integration. As our attention will be focused on behaviour as a function of time, this integration yields a performance criterion of the form

$$\int_0^T (\text{objective function}) \, dt$$

where the final time T may be infinity. This chapter outlines how such problems can be solved. Emphasis is placed on the basic ideas rather than on the precise details. More rigorous and complete accounts can be found in Athans and Falb (1966), Jacobs (1974) and Norman (1975).

The sort of problem we will be concerned with involves considering the animal to be choosing its behaviour so as to change its state in some desired way. This involves a change of emphasis compared to traditional approaches to motivation, which take motivational state as causing behaviour. The justification for the new viewpoint is the closed loop that links state and behaviour, as shown in Fig. 5.8. As a result, we can take either the motivational state or the behaviour to be fundamental.

The optimal control problem can new be defined. There is an objective function, K, dependent on the state, x, and the behavioural control, u. It is desired to move the system, perhaps for a specified time or to a specified state, such that the integral of the objective function is minimized. The classical techniques for solving this sort of problem assume that u does not change in a sudden way and has no upper limit. To the extent that behaviour changes suddenly (i.e. is discontinuous) and has a maximum rate (i.e. is constrained), such techniques cannot be applied. A technique that is applicable in such cases was developed by Pontryagin and his co-workers in the 1950's. We introduce this technique by way of the more general approach known as dynamic programming.

8.1 Dynamic programming

Dynamic programming is the name given to the theory of dynamic decision problems in which current choices may influence the options at a later time. The term 'programming' refers to the fact that such problems usually have to be solved by a computer program, and the basic equation of dynamic programming is well suited to this approach. Its main drawback is that it becomes impractical when the state has many dimensions. We give an example of the application of dynamic programming to behaviour in section 8.2.3.

8.1.1 The principle of optimality

The solution to dynamic optimization problems is based on the principle of optimality. This principle can be summarized as follows: 'An optimal policy has the property that, whatever the initial state and initial decision are, the remaining decisions must constitute an optimal policy with regard to the state resulting from the first decision,' (Bellman, 1957). The principle can be proved by contradiction by considering an optimal policy π^* from initial state $x(1)$ to final state $x(N)$ which involves going through state $x(2)$. Once $x(2)$ has been reached, π^* determines a policy from $x(2)$ to $x(N)$. If the principle of optimality did not hold, π^* would not be the optimal policy for this transition and so there would be some policy π that was better. Now, however, by following policy π^* from $x(1)$ to $x(2)$ and policy π from $x(2)$ to $x(N)$ we have constructed a policy which is better than π^* from $x(1)$ to $x(N)$. This contradicts the assumption that π^* was optimal, and therefore we must reject the assumption that the principle of optimality does not hold.

Another way of looking at the principle of optimality is that as long as the state provides a complete description of the decision problem, the way in which the present state was reached does not influence the optimal policy given this state. The fact is well known to good poker players, who do not persist with a hand just because they have already staked a large sum of money. The failure to realize that previous investment should not necessarily commit one to further investment has become known as the 'Concorde fallacy' because of the example used by Dawkins and Carlisle (1976). In analysing such an error in Trivers's (1972) treatment of parental investment they conclude: '... a parent "deciding" whether to desert a child should ask the following questions: How closely related is this child to me? How likely is it to survive if I do, and if I do not, desert it? What proportion of my total future reproductive potential does this child represent? How much would it cost me to make a new child equivalent to this one? The parent should ignore its own previous investment in the particular child, except insofar as it affects the answers to these questions'.

Although the principle of optimality may seem to be too simple to be useful, in fact it can be used to derive the most general recurrence relation for solving dynamic control problems. Some feel for how this is done can best be obtained by using discrete-time systems such as the automata described in Chapter 5. In this case the way in which the state changes is determined by the next-state function, δ. The state at time $i+1$ can therefore be written as

$$x(i+1) = \delta(x(i), u(i)) \tag{8.1}$$

(This involves a slight change of notation from that of Chapter 5. We are now considering the state, Q, to be represented by a single-state variable, x.) For some performance criterion V of the form

$$V = \sum_{i=1}^{N} C(x(i), u(i)) \tag{8.2}$$

a function $V_N^0(x)$ can be defined as follows: $V_N^0(x)$ is the value of V if the optimal policy is used for N steps starting from the initial state $x(1) = x$, i.e.,

$$V_N^0(x) = \min_{u(1),\ldots,u(N)} \{C(x, u(1), \ldots, u(N))\}$$

Consider the state $x(2)$ after the first decision $u(1)$ has been made. By the principle of optimality, the remaining $N-1$ steps constitute an

optimal policy for $x(2)$. Their cost or benefit is therefore given by:
$$V^0_{N-1}(x(2)) = V^0_{N-1}(\delta(x(1), u(1))),$$
as
$$x(2) = \delta(x(1), u(1))$$
This means that the total cost or benefit is
$$V = C(x(1), u(1)) + V^0_{N-1}(\delta(x(1), u(1)))$$
For the policy to be optimal, this expression must be minimized with respect to $u(1)$, which can be expressed as
$$V^0_N(x(1)) = \min_{u(1)} \{C(x(1), u(1)) + V^0_{N-1}(\delta(x(1), u(1)))\} \qquad (8.3)$$
which is the basic recurrence relation of dynamic control. This equation can be solved numerically by working back from when there is only one step left. For a continuous-time problem defined by
$$\dot{x} = f(x, u)$$
and
$$V = \int_0^T C(x, u) \, dt$$
equation (8.3) becomes
$$\frac{\partial V^0}{\partial T} = \min_u \left\{ C(x, u) + \frac{\partial V^0}{\partial x} f(x, u) \right\} \qquad (8.4)$$
(see Jacobs (1967) for details).

The vector version of equation (8.4) is
$$\frac{\partial V^0}{\partial T} = \min_u \left\{ C(\mathbf{x}, \mathbf{u}) + \sum_{j=1}^n \frac{\partial V^0}{\partial x_j} f_j(\mathbf{x}, \mathbf{u}) \right\}$$
replacing the time-to-go, T, by absolute time, t, yields:
$$-\frac{\partial V^0}{\partial t} = \min_u \left\{ C(\mathbf{x}, \mathbf{u}) + \sum_{j=1}^n \frac{\partial V^0}{\partial x_j} f_j(\mathbf{x}, \mathbf{u}) \right\} \qquad (8.5)$$
This is virtually the celebrated Pontryagin Maximum Principle, which we now describe.

8.2 Pontryagin's Maximum Principle

Pontryagin approached the optimal control problem by defining a state function, called the Pontryagin (or, sometimes, the Hamiltonian, see Appendix 8.1) and denoted by H. Pontryagin's Maximum Principle states that the problem of finding the path of least cost is equivalent to the more direct problem of instantaneously maximizing H. (The principle can also be written as an instantaneous minimization.) H is a function of state which depends on the plant equations and the objective function. The term 'plant equations' is taken from control theory and refers to the equations which characterize how behaviour determines the rate of change of state (i.e. the consequences of behaviour, as discussed in Chapter 5). The term is an apt one for the modelling of behaviour, because it underlines the idea of the animal controlling its internal state (i.e. a 'plant').

8.2.1 Costate variables

The Pontryagin function involves more than just the objective functions and the plant equations. The reason for this is a new sort of constraint, in addition to the limitations on the control vector. The nature of this constraint can be explained by comparing the static and dynamic optimization problems.

A general formulation of a static optimization problem that is exemplified in Chapter 7 can be characterized as follows:

Maximize (or minimize) some function of x_1 and x_2 subject to the constraint $p_1x_1 + p_2x_2 < M$ with $x_1 > 0$, $x_2 > 0$ (p_1 and p_2 are constants).

This sort of problem is well known in economics (Lancaster, 1968), where x_1 and x_2 represent goods and the function is interpreted as their utility. The prices of x_1 and x_2 are taken to be p_1 and p_2 respectively, so the constraint expresses the fact that expenditure, given by $p_1x_1 + p_2x_2$, cannot exceed available money, M.

As we saw in Chapter 7, one way to solve such problems is to introduce a function λ, known as a Lagrange multiplier. It is shown in Appendix 8.1 that λ is the change in optimal utility produced by a small relaxation of the constraint; in other words λ is the cost of the constraint (in economic theory λ is often known as the shadow price).

This static optimization problem takes account of the fact that x_1 and x_2 cannot be varied independently. In a similar way, the dynamic problem of optimal control must represent the fact that the terms involving x and the terms involving u that constitute the objective function cannot be varied independently. The reason for the dependence is the fact that u controls x; the nature of this control is given by the plant equations. In dynamic problems λ is a function of time, but its interpretation is still similar to that of the static case. In fact λ represents the change in total future cost along the optimal trajectory that results from a small change in state. (For details of the relationship between the static and dynamic problems, see Appendix 8.1.)

The Pontryagin function itself can be thought of as the gradient of the cost functional, that is to say H indicates how cost varies with chosen control at a given position and time.

Although Pontryagin's Maximum Principle is a powerful technique, it is not always possible to solve the equations that it yields. The major reason for this is that the state vector is known at the starting time but the costate vector is known at the final time. These split boundary conditions mean that it is not easy to integrate the dynamic equations.

Another drawback associated with Pontryagin's Maximum Principle is that it does not automatically express the optimal control as a function of state. Such a representation is known in systems theory as a control law (Schultz and Melsa, 1967) and in the theory of behaviour as a decision rule (McFarland, 1976). The application of Pontryagin's Maximum Principle results in an open-loop expression of the optimal control, whereas the idea of a decision rule implies a closed-loop formulation. It is possible to obtain the optimal control as a function of state by using the Hamilton–Jacobi equations (Schultz and Melsa, 1967), but this approach also presents difficulties. Because a whole range of initial conditions is being considered, the Hamilton–Jacobi equations are harder to solve than the Pontryagin equations.

We now summarize one formulation of Pontryagin's Maximum Principle: in order to minimize $\int_0^T C(\boldsymbol{x}, \boldsymbol{u}, t) \, dt$, $\boldsymbol{u}$ must be chosen in such a way as to instantaneously minimize

$$H = C(\boldsymbol{x}, \boldsymbol{u}, t) - \boldsymbol{\lambda}^* f(\boldsymbol{x}, \boldsymbol{u}, t) \tag{8.6}$$

where $\boldsymbol{\lambda}^*$ is the matrix transpose of the costate vector $\boldsymbol{\lambda}$, and $f(\boldsymbol{x}, \boldsymbol{u}, t)$ gives the plant equations, i.e.

$$\dot{\boldsymbol{x}} = f(\boldsymbol{x}, \boldsymbol{u}, t)$$

The costate is specified by the following equation for its rate of change:

$$\dot{\boldsymbol{\lambda}} = -\partial H / \partial \boldsymbol{x} \tag{8.7}$$

and there is a similar equation for the rate of change of the state:

$$\dot{\boldsymbol{x}} = \partial H / \partial \boldsymbol{\lambda} \tag{8.8}$$

It can be seen that the right-hand side of equation (8.6) is similar to the right-hand side of equation (8.5).

8.2.2 An example of the Pontryagin approach: satiation of feeding behaviour

The static and dynamic approaches can be contrasted by considering how an animal should feed. The approach often used in optimal foraging theory is to assume that the animal's goal is to maximize its rate of energy intake (see Krebs, 1978, for a review). If an animal starts with a given energy deficit which it must remove, the criterion of maximizing its rate of energy intake will mean that it feeds at a constant rate until the deficit reaches zero. This is obviously a static formulation of the problem. Dynamic optimization involves looking for a trajectory from the initial deficit along which the total value of the objective function is minimized. Sibly and McFarland (1976) suggest that the appropriate cost function is a quadratic, i.e.

$$C = a_1 x^2 + a_2 u^2 \tag{8.9}$$

where a_1 and a_2 are constants that 'weight' the relative importance of the cost associated with x and u, respectively. As we see below, although a_1 and a_2 are constant for any given problem, they may vary from problem to problem.

The cost function given in equation (8.9) can be simplified by rescaling a_1 relative to a_2. In other words, we set a_2 equal to unity and introduce a new constant, a^2, which is equal to a_1/a_2. The cost function is now given by

$$C = a^2 x^2 + u^2 \tag{8.10}$$

(The reason for using a^2 rather than a will become apparent when the optimal solution is given.)

The quadratic cost on x means that a given reduction in x reduces the cost by more when x is large than when it is small. The same relationship also holds between u and its cost. The idea of such a cost on x has been discussed in Chapter 7, but we have not yet commented on the cost of a given rate of behaviour. Both costs depend on a way in which an animal can endanger its life. The term $C(x)$ is based on the risk of the deficit crossing a lethal boundary. The term $C(u)$ reflects the danger of predation that arises from feeding at a rate u. It is assumed that an increase in feeding rate involves a reduction in the time spent looking around for potential predators and that the consequences of this reduction are more costly when u is high.

The optimization problem can now be stated. It is desired to minimize

$$\int_0^\infty (a^2 x^2 + u^2)\, dt$$

subject to the following conditions: (i) the deficit variable x starts at the value x_0; (ii) the plant equation is given by

$$\dot{x} = -ru;$$

and (iii) there is no effective upper limit on u.

Pontryagin's Maximum Principle can be used to derive the solution to this problem, as is shown in Appendix 8.2. (Indeed, condition (iii) means that classical techniques could also be used – see, for example Jacobs (1974), Chapter 9.) The optimal trajectory for the deficit is given by the following equation:

$$x(t) = x_0 e^{-rat} \tag{8.11}$$

which means that the control (i.e. behaviour) should have a trajectory of the form

$$u(t) = ax(t) \tag{8.12}$$

$$u(t) = ax_0 e^{-rat} \tag{8.13}$$

Equations (8.12) and (8.13) are equivalent forms of the control law for this problem. In equation (8.12) the control is expressed as a function of the state, whereas equation (8.13) gives the control as a function of time (recall that x_0 is constant for any given problem). The exponential decrease in the rate of behaviour can be justified on intuitive grounds. When x is big, it is worthwhile to feed rapidly to reduce x, because a change in x reduces the cost by a substantial amount. When x is smaller, the cost of a very high rate of behaviour would not be offset by the reduction in x, so u declines as x is reduced. The optimal behaviour results in the exponential decline in u given by equation (8.13).

The trajectory of the deficit can be converted to a satiation curve, in which cumulative intake is represented as a function of time spent feeding. If the optimal trajectory is given by equation (8.11), then the corresponding satiation curve $N(t)$ will be given by

$$N(t) = x_0 - x(t) = x_0 (1 - e^{-rat}) \tag{8.14}$$

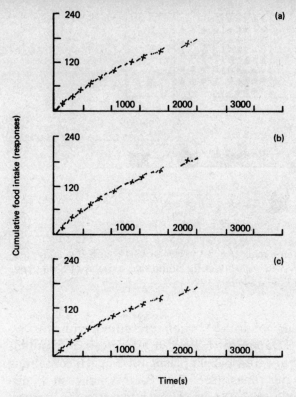

Fig. 8.1 Cumulative response data from a single rat showing fitted curves (+): (a) exponential, (b) parabola, (c) hyperbola. (After McCleery, 1977.)

This prediction about the form of the satiation curve was tested by McCleery (1977). Using the feeding behaviour of hungry rats, he fitted hyperbolas and parabolas as well as exponentials to the empirical satiation curves. These curves are illustrated in Fig. 8.1. Both the exponential and the hyperbola (of the form $N = At/(k+t)$, where A and k are constants) provided a good representation of the data in most cases. The analysis was then extended to take account of the 'warm-up' effect that was sometimes found. This involved removing the constraint that the fitted curve had to pass through the origin. This adds another parameter – the value of t when $N = 0$. This can be thought of as the time that elapses before responding 'really' starts. A comparison of the residual least-squares error suggested that the exponential was the best fit.

A similar approach to the problem of satiation can be found in Milinski and Heller (1978). They studied the behaviour of 3-spined

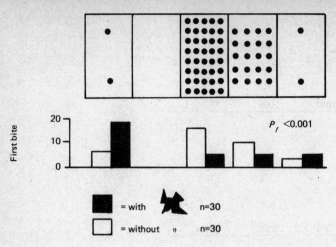

Fig. 8.2 Number of first bites made by 30 sticklebacks towards *Daphnia* prey presented at various densities, with or without the additional stimulus of a silhouette of a kingfisher. (After Milinski and Heller, 1978.)

sticklebacks (*Gasterosteus aculeatus*) when presented with various densities of water fleas (*Daphnia*). Each fish was given a simultaneous choice between a set of transparent plastic tubes, each containing a given number of water fleas (see Fig. 8.2). Under such circumstances the fish usually directed its initial attacks to the densest swarm. If, however, the silhouette of a kingfisher was glided over the fish before the experiment began, it was found that the initial attacks were delivered to the less-dense swarms.

Milinski and Heller account for this result by postulating the following cost function

$$C = \alpha x^2 + \beta w u^2 \qquad (8.15)$$

where α and β are weighting parameters, w is the density of water fleas, and u is the fish's capture rate.

Milinski and Heller suggest that high capture rates can be achieved by attacking dense swarms, but this increases the stickleback's risk of predation. Their justification for this claim is that large swarms subject the stickleback to the so-called 'confusion effect' (Bertram, 1978; Neill and Cullen, 1974), and to overcome this the fish has to concentrate its attention on the water fleas to the exclusion of being aware of potential predators. As a result, the cost depends on both the density, w, and the capture rate, u.

The optimal choice of capture rate and prey density for a stickleback with an initial food deficit x_0 can be found by minimizing

$$\int_0^T \alpha x^2 + \beta w u^2 \, dt \quad \text{subject to} \quad \dot{x} = -ru.$$

Using Pontryagin's Maximum Principle to solve this problem numerically, Milinski and Heller (1978) report that the capture rate and the density in which the fish hunts should fall as the fish becomes satiated. Gliding the silhouette of the kingfisher over a stickleback should increase the stickleback's assessment of the risk from predation, which amounts to increasing the weighting parameter β (see also Chapters 9 and 10). The effect of an increase in β on the optimal behaviour sequence will be to reduce the initial capture rate.

There are two problems with this work, both of which stem from the fact that the sticklebacks are not able to consume the water fleas. Firstly, it is not possible to check the assumption about capture rate as a function of density. Secondly, although the model predicts the time-course of behaviour to satiation, it is only really possible to observe the initial choice of density. Milinski and Heller claim more than this by identifying satiation with the habituation of the attack response, but this is not completely convincing. Subsequent work (Heller and Milinski, 1979) has overcome these problems. There are now data on the capture rate of sticklebacks as they approach satiation, but this has not been combined with the use of the kingfisher silhouette in an attempt to manipulate β.

Heller and Milinski (1979) present the optimal behaviour for both variable and constant prey density and also consider the case of an upper limit on capture rate. We will confine our discussion of their results to a fish feeding at a rate below the limit in prey of constant density, because this case can be easily compared with the quadratic formulation (equation (8.9)). We use the comparison to explain what might at first sight be a paradoxical prediction about the effect of prey density on the optimal solution to such cases.

If prey density, w, is kept constant, equation (8.15) becomes formally identical to equation (8.9). By identifying $\alpha/\beta w$ with a^2, the optimal solution to the quadratic case can be used to obtain the optimal capture rate. We take equation (8.13) and replace a by $\sqrt{\alpha/\beta w}$:

$$u(t) = \sqrt{\alpha/\beta w} \cdot x_0 e^{-rt\sqrt{\alpha/\beta w}} \tag{8.16}$$

At the start (i.e. when $t=0$) the exponential term is equal to unity and so equation (8.16) reduces to

$$u(0) = \sqrt{\alpha/\beta w} \cdot x_0 \tag{8.17}$$

(This also follows directly from equation (8.12).)

Equation (8.17) indicates that, for a given value of initial food deficit x_0, the optimal capture rate in a low density is higher than the optimal capture rate in a high density (see Heller and Milinski, 1979, p. 1133). Because the prey density enters into the cost, reducing density reduces the cost of a given capture rate and so the optimal capture rate increases. Heller and Milinski (1979) present evidence for this relationship between capture rate and density. The time required for the least hungry sticklebacks ('Group 4') to eat the first 8 water fleas was longer when density was high that when it was low. In other words, the rate was less in the high-density condition. (Data from the least hungry fish are used in the hope that the capture rate will not be constrained by the upper limit.)

Cowie *et al.* (1981) report some experiments which consider the complete course of feeding to satiation in an attempt to demonstrate the effects of the risk of being attacked by a predator. Hungry great tits were given meal worms by means of the conveyor belt described by Krebs *et al.* (1977). In one experiment the birds were tested under 3 conditions, each condition lasting for a week. The control condition involved letting the birds feed to satiation without any disturbance. In the 'hawk' condition, the great tit was shown a stuffed sparrowhawk just before the start of the test. The sparrowhawk (*Accipiter nisus*) is known to be a major predator on great tits (Perrins, 1979) and all the birds showed signs of fright when presented with the stuffed sparrowhawk. The 'torch' condition was similar except that the great tit was shown a torch instead of a sparrowhawk. The order of the conditions was randomized across the birds. In the 'torch' condition the time taken to eat an item, the time between items and the frequency of 'look ups' all increased with cumulative intake, as is shown in Fig. 8.3. A comparison of the 'hawk' and 'torch' conditions revealed an increase in the first two measures after the presentation of the hawk.

A dynamic approach to foraging behaviour over a longer time period is discussed by Katz (1974). He proposed an optimality principle for the feeding behaviour and associated weight changes of the weaver bird *Quelea quelea*. The optimality criterion is to minimize the time spent foraging during the year while avoiding

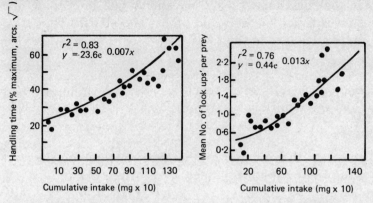

Fig. 8.3 Two measures of the response of great tits to briefly presented prey. (After Cowie et al., 1981.)

death from starvation. The resulting model shows a good qualitative agreement with the data of Ward (1965).

A similar optimality criterion has been applied by Craig, De Angelis and Dixon, (1979) to the foraging behaviour of the loggerhead shrike (*Lanius ludovicianus gambeli* Ridgway). Their long-term model (over 10 months of the year) predicts attack rates which show the same general trend as the data, but the fit is not very good. This may result from extrapolations in the model concerning prey availability. There was also some indication that a function of time feeding and body weight would provide a better optimality criterion. As the authors say: 'It is probable that, although the adult shrike can tolerate weights down to 45 g or less, it runs a somewhat higher risk of mortality at that level than at a weight of 50 g, which is the approximate weight of adult shrikes. It may be advantageous for the shrike to trade a certain amount of extra time devoted to feeding..... in order to keep its weight closer to the ideal of 50 g.'

Regardless of the exact fit of such models, their main virtue is that they give an account of seasonal variations in feeding behaviour, and are therefore able to predict anticipatory strategies. For example, because of prey shortage in November and December, the model puts on weight during October. This is a testable prediction.

8.2.3 An example of dynamic programming – feeding and drinking with delays between changes

The discussion of satiation has so far confined itself to the simple problem of one unconstrained activity. The case of a hungry and thirsty animal that has limits on the maximum rate of feeding and drinking was investigated by Sibly and McFarland (1976). They state that if the cost function is of the form

$$C = x_1^2 + x_2^2 + u_1^2 + u_2^2, \tag{8.18}$$

where x_1 = food deficit, x_2 = water deficit, u_1 = rate of eating, u_2 = rate of drinking, and the plant equations are

$$\dot{x}_i = -r_i u_i \qquad i = 1, 2$$

then the behaviour that minimizes the time integral of the cost function is given by the following rule:

eat if $\quad x_1 r_1 k_1 > x_2 r_2 k_2$

drink if $\quad x_2 r_2 k_2 > x_1 r_1 k_2,$

where k_1 and k_2 are the upper limits on u_1 and u_2, respectively. This means that there is a switching line given by $x_1 = x_2 \dfrac{r_2 k_2}{r_1 k_1}$, on one side of which the animal should feed and on the other side of which it should drink.

The problem is harder to solve if there is a cost of changing from one activity to the other (see Chapter 6 and Larkin and McFarland, 1978). Larkin (1981) uses dynamic programming to tackle this problem. We will only discuss the case in which the cost is a time delay. This delay increases the total cost of reaching satiation, because the deficits cannot be reduced during it. Larkin assumed the quadratic cost function given by equation (8.9), but it was not possible in practice to vary the rates of the activities as well as the pattern of alternation between them. Larkin therefore held u_1 and u_2 fixed at their maximum value. It is very likely that this is the optimal strategy unless the deficits are small. Alternatively, we can consider the objective function to be

$$C = A_1 x_1^2 + A_2 x_2^2 \quad (A_1, A_2 \text{ are constants}),$$

in which case the animal should always feed or drink at the maximum possible rate, and the problem becomes one of finding the sequence of feeding and drinking that is least costly.

As would be expected from a dynamic programming algorithm, Larkin's procedure is to work backwards from the satiation point (i.e., $x_1 = x_2 = 0$). First of all, the minimum cost of reducing x_1 from 1 to zero is calculated, given that $x_2 = 0$ (assuming that everything is scaled so that each decision to feed reduces the appropriate deficit by one unit. Note that the problem has now been made discrete, as compared to the continuous case considered by Sibly and McFarland (1976)). There is only one way to get to $x_1 = x_2 = 0$ from $x_1 = 1$, $x_2 = 0$, so calculating the cost of this step gives the minimum cost. Starting from $x_1 = 2$, $x_2 = 0$ adds another step (again with no choice) and hence another cost. In this way the minimum cost for all points along the line $x_2 = 0$ can be calculated. Points along the line $x_2 = 1$ are then evaluated under two conditions: (1) the animal is already feeding; (2) the animal is already drinking. For example, at the point $x_1 = x_2 = 1$, in condition 1 the choice is between (a) continuing feeding, bringing x_1 to zero and then switching to drinking and, after the delay during which x_2 stays at one, reducing x_2 to zero, and (b) switching to drinking, with a delay during which x_1 and x_2 are equal to one, then reducing x_2 to zero, switching back to feeding with a delay during which x_1 stays on one, and then reducing x_1 to zero (for which the minimum cost has already been found). Comparing (a) and (b) yields the minimum cost of getting from $x_1 = x_2 = 1$ to $x_1 = x_2 = 0$, and this can be used, together with the minimum cost for $x_1 = 2$, $x_2 = 0$, to find the minimum cost for $x_1 = 2$, $x_2 = 1$. Repeating this procedure for bigger and bigger values of x eventually produces the minimum cost of reaching satiation from every point within a finite set of possible initial deficits. The optimal behaviour can be specified

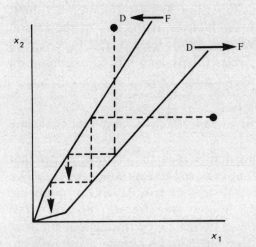

Fig. 8.4 The general form of optimal policy when there are switching costs. The line marked D←F determines transitions from feeding to drinking, and the line marked D→F determines transitions from drinking to feeding. Two possible trajectories are illustrated showing that the behaviour between the switching lines is not completely determined by the values of x_1 and x_2.

by two switching lines, as opposed to the single line $x_1 r_1 k_1 = x_2 r_2 k_2$ that represents the solution when there are no costs of changing. The optimal policy depends on which activity is currently being performed, so one line determines transitions from feeding to drinking, and the other determines transitions from drinking to feeding. To the left of the feeding to drinking boundary the optimal behaviour is to switch to drinking, and to the right of the drinking to feeding boundary, it is to switch to feeding. Some trajectories illustrating these rules are given in Fig. 8.4.

It has been assumed in this discussion that the delay occurs following a transition in either direction, but it is obviously straightforward to find the switching lines when the delay only occurs after a transition in one direction. Even in this case it turns out that both switching lines are different from the switching line for the same problem without the delays.

8.2.4 The reproductive behaviour of social insects

Our next example of dynamic optimization provides a link between the short-term optimization (such as satiation effects) and the question of optimal life histories. The problem to be discussed is how a colony of eusocial insects should allocate their resources between workers and reproductives (i.e. queens and males). Macevicz and Oster (1976) assume that in competition between annual colonies, the colony that has the most reproductives at the end of the season will be at an advantage. (See Wilson, 1975, for a detailed account of insect colonies and their life cycles.) This amounts to assuming an optimality criterion of the form

$$\text{maximize } J = Q(T),$$

where Q represents the number of reproductives and T is final time (i.e., the end of the season).

The control variable is proportion $u(t)$ of the colony's effort that goes to producing workers (W) as opposed to reproductives (Q). The resulting dynamic equations for $\dot{W}$ and $\dot{Q}$ will depend on both the size of the workforce (i.e. W) and on the 'return function' $R(t)$ which characterizes the abundance of resources throughout the season and the insects' ability to harvest them. A simple assumption to make is that production is linear to both W and $R(t)$, i.e.,

$$\text{Production} = bu W R(t).$$

The expression for $\dot{W}$ is given by production minus mortality, so if probability of death is a constant μ_1, then one dynamic equation would be

$$\dot{W} = buWR(t) - \mu_1 W \qquad (8.19)$$

The corresponding equation for Q would be

$$\dot{Q} = c(1-u)WR(t) - \mu_2 Q \qquad (8.20)$$

As the optimality criterion $J = Q(T)$ can be rewritten as

$$J = \int_0^T \dot{Q}\, \mathrm{d}t, \qquad C = \dot{Q},$$

and so the Pontryagin function takes the form

$$H = \lambda_1 \dot{W} + \lambda_2 \dot{Q} - \dot{Q} \qquad (8.21)$$

By instantaneously maximizing this function, Macevicz and Oster (1976) found that the optimal policy involves u taking only the values zero and one, which is known as a 'bang-bang control'. In other words the colony should devote all its effort to the production of workers until near the end of season, when it switches to devoting all its effort to reproductives. This result does not depend on the form of the returns function. For some particular cases the switch occurs as late as possible, i.e., when there is only one mean worker lifetime left. This sort of strategy has been observed for the wasp *Vespa orientalis* by Ishay, Bytinski-Salz and Shulov (1967) and the hornet *Polistes fuscatus* by Eberhard (1969).

Although the bang-bang strategy is robust in the face of changes in the return function, it may not hold when the dynamic equations show diminishing returns. The reader should consult Macevicz and Oster (1976) and Oster and Wilson (1978) for further details. The latter reference also discusses colonies that reproduce over several seasons.

For an annual colony, the problem of allocating a season's resources between workers and reproductives amounts to determining the optimal life history for the insects that belong to the colony. The subject of optimal life histories is often considered to be part of ecology rather than of animal behaviour. We believe this to be an artificial distinction, encouraged by the formulation of life history strategies in terms of growth versus reproduction. A more flexible approach to life histories is developed by Freeman and McFarland (1981), based on Fisher's (1930) age-specific reproductive potential. In the next section we look at the argument in more detail.

8.3 Optimal life histories

Sibly and McFarland (1976) defined the cost per unit time as the death rate minus the birth rate, i.e.,

$$\frac{\text{Cost}}{\text{Time}} = m(\boldsymbol{x}, \boldsymbol{u}) - r(\boldsymbol{x}, \boldsymbol{u}) \tag{8.22}$$

where $m(\boldsymbol{x}, \boldsymbol{u})$ = probability of the animal dying in unit time

$r(\boldsymbol{x}, \boldsymbol{u})$ = probability of the animal giving birth in unit time.

They claimed that minimizing the integral of this function over time would give the trajectory that maximized the animal's fitness. However, subtracting a birth rate from a death rate is not really valid, and it fails to take account of the dependence of reproductive potential on survival. Freeman and McFarland (1981) develop the argument from first principles and show that the death rate minus birth rate formulation is only correct when one of the rates is zero. It is easy to get an intuitive idea of the correct formulation of the optimization problem. Fitness has to do with reproduction, and the importance of mortality is that it curtails reproduction. There is no intrinsic advantage of survival; the advantage of surviving is that it may permit further reproduction. The essence of the animal's problem is to decide between reproduction now and reproduction in the future. Both the value of future reproduction and the probability of surviving to achieve it depend on the animal's choice of activities. The probability of surviving can be represented by an integral of the mortality rate (which depends on $\boldsymbol{x}$ and $\boldsymbol{u}$), and so the function to be integrated itself contains an integration. With some looseness of notation, the argument can be written as follows:

Probability of being alive at time t' if alive at time t is

$$\exp\left(-\int_t^{t'} m(\boldsymbol{x}, \boldsymbol{u}) \, \mathrm{d}t\right) \tag{8.23}$$

and so the fitness is proportional to the integral from t to infinity of

$$r(\boldsymbol{x}, \boldsymbol{u}) \exp\left(-\int_t^{t'} m(\boldsymbol{x}, \boldsymbol{u}) \, \mathrm{d}t\right) \tag{8.24}$$

It can be seen that equation (8.24) is substantially different from equation (8.22). For ease of reference we express both forms as maximization problems.

$$\text{Max} \int_0^\infty r(\mathbf{x}, \mathbf{u}) - m(\mathbf{x}, \mathbf{u}) \, dt \tag{8.25}$$

$$\text{Max} \int_0^\infty r(\mathbf{x}, \mathbf{u}) \exp\left(-\int_t^{t'} m(\mathbf{x}, \mathbf{u}) \, dt\right) \tag{8.26}$$

Freeman and McFarland (1981) show that the problem of choosing $\mathbf{u}$ to maximize equation (8.26) can be solved by choosing $\mathbf{u}$ so as instantaneously to maximize

$$r(\mathbf{x}, \mathbf{u}) - m(\mathbf{x}, \mathbf{u})\hat{R} + \sum_{i=1}^n \frac{\partial R}{\partial x_i} f_i(\mathbf{x}, \mathbf{u}) \tag{8.27}$$

where $\hat{R}$ is an index of future reproductive potential, and

$$f_i(\mathbf{x}, \mathbf{u}) = \dot{x}_i \tag{8.28}$$

The corresponding immediate maximization for the Sibly and McFarland formulation is

$$r(\mathbf{x}, \mathbf{u}) - m(\mathbf{x}, \mathbf{u}) + \sum_{i=1}^n \frac{\partial V^0}{\partial x_i} f_i(\mathbf{x}, \mathbf{u}) \tag{8.29}$$

where $\frac{\partial V^0}{\partial x_i} = \lambda_i$, and V^0 is the integral of the objective function along the optimal trajectory (see Appendix 8.1).

The Sibly and McFarland formulation is strictly incorrect in directly subtracting terms relating to reproduction and mortality which are not strictly comparable. (The Freeman and McFarland formulation effectively weights mortality by an index of future reproductive potential.) However, this formulation can be applied in situations which focus on reproductive behaviour, so that mortality can be neglected, or which concentrate on survival strategy and neglect reproduction. This is the case with our use of the Sibly–McFarland formulation in sections 8.2.2 and 8.2.3.

The Freeman–McFarland formulation implies that the objective function, parametized by $\hat{R}$ in the simpler cases, changes throughout the animal's life history. It is therefore likely that the goal function will undergo changes during ontogeny, particularly at the time when the animal becomes sexually active, and perhaps even during senescence. We can expect that some of these changes will be genetically preprogrammed and some will be due to learning. However, as we argue in Chapter 10, only preprogrammed types of learning, such as imprinting, can lead to alterations in the goal function.

8.4 Summary

Dynamic optimization problems are concerned with situations in which the action taken at any time has consequences which are evaluated over some period in the future. The problem is to find a set of design parameters, which are all continuous functions of some other parameter, which itself minimizes an objective function subject to the prescribed constraints. Such problems can be solved by means of dynamic programming, or by the application of Pontryagin's Maximum Principle. Both approaches have been used in animal behaviour and examples are discussed.

Appendix 8.1

Pontryagin's Maximum Principle and the costate variables

Pontryagin's Maximum Principle states that to minimize $\int_0^T C \, dt$ for some cost function C, $\mathbf{u}$ must be chosen so as to maximize the Pontryagin state function, H:

$$H = \lambda^* f(\mathbf{x}, \mathbf{u}, t) - C(\mathbf{x}, \mathbf{u}, t) \tag{1}$$

where λ^* is the matrix transpose of the costate or adjoint vector λ and $\dot{\mathbf{x}} = f(\mathbf{x}, \mathbf{u}, t)$. The costate vector is, in effect, a set of Lagrange multipliers, introduced to satisfy the plant-equation constraints. It is given by the equation

$$\dot{\lambda} = -\partial H / \partial \mathbf{x}$$

The same results can also be achieved by a minimization principle. The optimum control is the one that minimizes a slightly different Pontryagin, $\tilde{H}$, which is defined as

$$\tilde{H} = C(\mathbf{x}, \mathbf{u}, t) + \lambda^* f(\mathbf{x}, \mathbf{u}, t)$$

where

$$\dot{\lambda} = -\partial \tilde{H} / \partial \mathbf{x}$$

The equivalence of these two principles follows from the fact that the maximum of a given function is equal to the minimum of the negative of the function. This means that maximizing

$$H = \boldsymbol{\lambda}^* \dot{\boldsymbol{x}} - C$$

is equivalent to minimizing

$$-H = C - \boldsymbol{\lambda}^* \dot{\boldsymbol{x}}$$

which is equivalent to minimizing

$$\tilde{H} = C + \boldsymbol{\lambda}^* \dot{\boldsymbol{x}}$$

if $\dot{\boldsymbol{\lambda}}$ is defined as $-\partial \tilde{H}/\partial \boldsymbol{x}$ rather than $-\partial H/\partial \boldsymbol{x}$.

In some circumstances equation (1) is formally identical to the Hamiltonian of classical mechanics (Rosen, 1970, gives a clear account of the Hamiltonian formulation of equations of motion), but the two functions are different. The variables in the Hamiltonian that correspond to the costate in Pontryagin's function are momentum variables. The meaning of the costate variables is totally different. Its explanation requires a new function, V^0, defined as follows:

$$V^0 = \int_t^T C(\boldsymbol{x}, \boldsymbol{u}^0, t) \, dt$$

where $\boldsymbol{u}^0$ is the optimal control policy. V can therefore be thought of as the total value of the objective function along the optimal trajectory. The costate vector is the change in V^0 that results from a small change in state, i.e.

$$\boldsymbol{\lambda} = \partial V^0 / \partial \boldsymbol{x}$$

Another way to present the costate variable is to illustrate the role of $\boldsymbol{\lambda}$ in a static optimality problem and then to consider the relationship between the static and the dynamic cases. The example given can be thought of in terms of which combination of commodities to buy (Chapter 6) or which behaviour to give up (Chapter 7).

The Static optimization problem

Given the problem of maximizing some function $B(\boldsymbol{x})$ subject to the constraint $d(\boldsymbol{x}) \geq 0$, the optimal solution can be found by constructing the function $L(\boldsymbol{x}, \boldsymbol{\lambda})$:

$$L(\boldsymbol{x}, \boldsymbol{\lambda}) = B(\boldsymbol{x}) + \boldsymbol{\lambda} \, d(\boldsymbol{x})$$

The Kuhn–Tucker conditions state that $B(\boldsymbol{x})$ is maximized by finding the saddle point of $L(x, \lambda)$, i.e. by maximizing with respect to x and minimizing with respect to λ. (See Lancaster (1968) for further details.)

For the purpose of this argument, assume that there are two state variables x_1 and x_2, and that the constraint function $d(x)$, is given by

$$p_1 x_1 + p_2 x_2 \leq M \tag{2}$$

In many problems of practical importance, the optimal solution lies on the equation of constraint, i.e. equality holds in equation (2). Under such circumstances, the constraint equation can be written as

$$M - p_1 x_1 - p_2 x_2 = 0$$

$L(x, \lambda)$ can therefore be written as

$$L(x, \lambda) = B(x_1, x_2) + \lambda (M - p_1 x_1 - p_2 x_2)$$

Differentiating with respect to the states and equating to zero yields

$$\frac{\partial B}{\partial x_1} - \lambda p_1 = 0 \qquad \frac{\partial B}{\partial x_2} - \lambda p_2 = 0$$

$$\therefore \quad \frac{\partial B}{\partial x_i} = \lambda p_i \qquad i = 1, 2$$

$$\therefore \quad \frac{\partial B}{\partial x_i} dx_i = \lambda p_i \, dx_i \qquad i = 1, 2$$

$$\therefore \quad \frac{\partial B}{\partial x_1} dx_1 + \frac{\partial B}{\partial x_2} dx_2 = \lambda (p_1 \, dx_1 + p_2 \, dx_2) \tag{3}$$

But the left-hand side of equation (3) is dB, and as $M = p_1 x_1 + p_2 x_2$, the term in brackets is dM

$$\therefore \quad dB = \lambda \, dM$$

Rearranging and taking to the limit:

$$\lambda = \frac{\partial B}{\partial M}$$

i.e. λ is increase in utility produced by an increase in budget.

The relationship between the static and dynamic cases is summarized in Table 8.1. The analogy with economic theory suggests an

Table 8.1 Relationship between static and dynamic cases

Nature of problem	Static	Dynamic
Function to be maximized	$B(x)$	$\int_0^T -C(x)$
Equation of constraint	$d(x) \geq 0$	$\dot{x} = f(x, u)$
New function	$L(x, \lambda) = B(x) + \lambda\, d(x)$	$H(x, u, \lambda) = C(x, u) + \lambda f(x, u)$
Alternative method of solution	Use each constraint to eliminate a state variable.	Solve plant equations and hence eliminate u; work with $\dot{x}$ and x.
Interpretation of Lagrange multiplier	Change in B at optimum produced by a small change in the constraint.	Change in integral of C along optimal trajectory from present position as a result of a small change in state.

interpretation of λ that may be useful in the study of adaptation. In the static problem, λ represents the cost of the constraint. If λ equals zero, then relaxing the constraint does not change the value of the function B (which can be thought of as 'benefit'). In the same sort of way, λ_i in dynamic problems indicates the effect of the state x_i on the total future fitness of the animal.

Appendix 8.2

We seek the behaviour that minimizes

$$\int_0^\infty (a^2 x^2 + u^2)\, dt \qquad (1)$$

where

$$\dot{x} = -ru$$

Define

$$H = \lambda \dot{x} - C(x, u) \qquad (2)$$

Pontryagin's maximum principle says that (1) is minimized by choosing u so as instantaneously to minimize (2). In this case we can write H as an explicit function of u and then differentiate with respect to u and equate the resulting expression to zero:

$$H = -\lambda r u - u^2 - a^2 x^2$$

$$\frac{\partial H}{\partial u} = -\lambda r - 2u$$

$$\frac{\partial H}{\partial u} = 0 \quad \text{implies} \quad u = \frac{-\lambda r}{2}$$

$$\therefore \quad H(\text{opt}) = \frac{\lambda^2 r^2}{2} - \frac{\lambda^2 r^2}{4} - a^2 x^2$$

$$= \frac{\lambda^2 r^2}{4} - a^2 x^2$$

By definition,

$$\dot{\lambda} = \frac{-\partial H}{\partial x} = 2a^2 x \tag{3}$$

and

$$\dot{x} = \frac{\partial H}{\partial \lambda} = \frac{\lambda r^2}{2} \tag{4}$$

Differentiation of equation (4) gives

$$\ddot{x} = \frac{\dot{\lambda} r^2}{2}$$

$$= r^2 a^2 x \quad \text{(by substitution from equation (3))}. \tag{5}$$

The general solution of equation (5) is

$$x(t) = A e^{-rat} + B e^{rat} \tag{6}$$

If we require $x \to 0$ as $t \to \infty$, then $B = 0$. The other boundary condition means that $A = x(0)$, i.e. $x(t) = x(0) e^{-rat}$.

9 Goal functions and objective functions

The previous two chapters have described the way in which optimal behaviour can be derived from the minimization of objective functions. It will be recalled from section 6.2.2, however, that we must be careful to distinguish between an animal optimizing with respect to internal criteria (the goal function) and an animal behaving optimally with respect to natural selection (cost function). Establishing the cost function requires careful fieldwork and will not be considered here (see McFarland, 1977; McCleery, 1978, for a discussion). Our main concern in this chapter is the relationship between objective functions and goal functions.

An animal may appear to be behaving optimally if its behaviour conforms to a specified objective function, but this does not necessarily mean that it is behaving optimally with respect to a particular goal (or cost) function. To understand this we have to look carefully at the inverse optimality approach as a means of obtaining objective functions. This approach is mathematical rather than biological in that it attempts to construct the objective function from a description of the behaviour without any preconceptions about the costs and benefits involved. In section 9.1.1 we discuss some technical aspects of this procedure, which is then illustrated in section 9.1.2. Section 9.2 considers possible ways in which animals could achieve near optimal behaviour by employing decision rules, and section 9.3 discusses ways in which the rules might adapt to changing circumstances.

9.1 The inverse optimality approach

9.1.1 Some general remarks on inverse optimality

The inverse optimality approach assumes that an optimality principle underlies the behaviour of the animal and attempts to discover the form of the objective function from the observed behaviour.

Inverse optimality does not seem to have received much attention as a mathematical problem, so our conclusions may have to be revised in the light of thorough analysis. There are two points that we wish to raise. The first concerns the relationship between the objective function and the goal function. Just because an objective function has been found by inverse optimality, does not necessarily mean that the goal function has been found. The difficulties that surround the interpretation of objective functions are underlined by the fact that physical systems can be described in terms of the minimization or maximization of a function. The term 'extremal principle' is used to refer to such representation. Hamilton (1834) recast the equations of Newtonian dynamics into a form in which the path of a particle is represented as a path along which a certain mathematically defined quantity is smaller than along any other path. Other extremal principles are discussed by Born (1939) and Rosen (1967). Extremal principles were formerly attributed to divine design, but are now seen as examples of our ingenuity in describing the universe. In the words of Born (1939): 'It is not nature that is economical but science.' If an extremal principle is found for an animal's behaviour, it may be no more than such an economical description. When we believe this to be the case, or when we do not wish to be specific, we will use the term 'objective function'. Objective functions will be denoted by K, in contrast to goal functions, G, and cost functions, C. In principle the mathematical procedure of deriving an extremal principle from the observed behaviour may result in an objective function rather than a goal function. In practice, it may be impossible to tell the difference between the two.

The second problem is that the inverse optimality procedure, regarded as a mapping from behaviour (i.e. trajectories in state space) to functions, may not be continuous. Loosely speaking, a mapping is said to be continuous if objects that are in some sense close to each other are mapped to objects that are also close. The exact details do not concern us here, for the term will only be used as a convenient shorthand for the surprising change in optimality function that can result from a small change in state trajectory. A good illustration of such a change is provided by the work of McCleery (1977). As was described in Chapter 8, he tried to discriminate between three putative forms of satiation curve. As well as performing a statistical analysis of goodness of fit, McCleery also suggested that it would be instructive to find the function that each of the trajectories minimizes. One of the satiation curves suggested by McCleery is a power function of the form

$$N(t) = At^k$$

where $N(t)$ is the cumulative intake up to time t and A and k are constants. For the curve to decelerate, k must be greater than zero and less than one. Houston (unpublished) shows that this satiation curve minimizes the integral of an objective function of the form

$$K_1 = ABk^{k-1}N^{k-1} + \frac{BN^k}{k(k-1)} \tag{9.1}$$

where B is a constant.

The cumulative intake N is given by the initial deficit, X, minus the current deficit, x, i.e.

$$N = X - x$$

$$\therefore \dot{N} = -\dot{x}$$

$$= ru.$$

Substituting for N and $\dot{N}$ in equation (9.1) gives

$$K_1 = AB^{k-1}(X-x)^{k-1} + \frac{Br^k u^k}{k(k-1)} \tag{9.2}$$

The objective function given by equation (9.2) does not look very similar to the quadratic function (repeated here as equation (9.3)) that generates exponential satiation curves (see Chapter 8).

$$K_2 = a_1 x^2 + a_2 u^2 \tag{9.3}$$

The difference between K_1 and K_2 can be underlined by noting that exponentials with different time constants can be generated by changing the magnitude of a_1 or a_2 in equation (9.3), whereas a change in k for the parabola requires an alteration of the power to which x and u are raised in equation (9.2).

Despite the differences between K_1 and K_2, they are similar in that their associated optimal trajectories are very much alike for a certain range of values. We take this as an indication of the importance of distinguishing between an optimality principle as a description of behaviour and as a representation of the relationship between behaviour and fitness. We use the general term 'objective function' in the former context and reserve the terms 'goal function' and 'cost function' for the latter (see section 6.2.1).

9.1.2 Example – courtship behaviour of newts

The courtship behaviour of the male smooth newt has already been described in Chapter 5. In the same chapter we presented a causal model, called NEWTSEX IV, that behaved in the same way as male newts. The functional model proposed by Dempster (see McFarland, 1977; McCleery, 1978) can be summarized as follows:

The male is assumed to be designed to maximize the probability of successful fertilization. It is claimed that this probability* is given by

$$\int_0^\infty (e^{-\phi St} Su_1 x + Su_3 x)\, dt \tag{9.4}$$

where $S =$ the number of spermatophores that can be deposited, $x =$ oxygen need, S and x are the state variables. The full set of control variables is $u_1 =$ rate of display, $u_2 =$ rate of creep, $u_3 =$ rate of spermatophore transfer, and ϕ is the discount factor associated with display. The plant equation is

$$\dot{x} = \theta - x^{-\rho} u_2$$

θ is thus the basic rate at which oxygen need increases, but during creep this rate is decreased by an amount that depends on the current value of the oxygen need. The equation reflects the idea that creep is probably the least energetic part of the courtship sequence.

The equation for the Pontryagin function, H, is

$$H = e^{-\phi St} Su_1 x + Su_3 x - \lambda(\theta - x^{-\rho} u_2) \tag{9.5}$$

In this formulation the rate of change of the costate is given by $\partial H/\partial x$, which means that

$$\dot{\lambda} = e^{-\phi St} Su_1 + Su_3 - \rho \lambda x^{-(\rho+1)} u_2 \tag{9.6}$$

Before discussing this model, we will describe its derivation. The starting point was the causal model of the male newt's courtship behaviour known as NEWTSEX IV. A general account of NEWTSEX IV is given in Chapter 5, so we will confine our attention here to the part of the model that is the most relevant to the optimization model – the behaviour of the variable called hope. Hope, h, starts at a value that is proportional to S and then rises at a rate that increases with both display rate, u_1, and oxygen need, x. Display rate is itself dependent on S, so the rate of change of hope is proportional to S and x. In the model, the dependence is of the form

$$\dot{h} = r_1 Sx \tag{9.7}$$

where r_1 is a scaling constant.

* But note that the value of equation (9.4) does not have to lie between 0 and 1.

During creep, hope falls at a rate that decreases with increasing oxygen need. NEWTSEX IV was run with dynamics of the form

$$\dot{h} = (-r_2 h)/x \qquad (9.8)$$

and

$$\dot{h} = -r_3 h(r_4 - x) \qquad (9.9)$$

where r_2, r_3 and r_4 are scaling constants. (The published results are from the models using equation (9.9).)

It was Dempster's view that hope had no physiological reality and so could not be a state variable. He proposed that x was the state variable and h was the costate variable. On the assumption that the costate behaved in the same way during spermatophore transfer as it did during display, Dempster suggested the following equation for the rate of change of the costate:

$$\dot{\lambda} = Su_1 - \rho \lambda x^{-(\rho+1)} u_2 + Su_3 \qquad (9.10)$$

Equation (9.10) incorporates something like equation (9.8) for the decay during creep.

As $\lambda = \partial H/\partial x$, equation (9.10) means that the terms in H that depend on x during u_1 and u_3 are Sxu_1 and Sxu_3, respectively. The relative magnitude of these terms depends only on u_1 and u_3, so that if $u_1 > u_3$, the model will never deposit a spermatophore, because the display term will always be bigger than the transfer term. To overcome this problem, Dempster introduced a discount factor, $e^{-\phi St}$, that devalued display as a function of the time t that had elapsed since the start of courtship. As a result, the equation for $\dot{\lambda}$ is given by equation (9.6) instead of equation (9.10). Integrating equation (9.6) with respect to x yields those terms in H which depend on x:

$$H_x = e^{-\phi St} Su_1 x + Su_3 x + \lambda x^{-\rho} u_2 \qquad (9.11)$$

The presence of λ indicates that the term $x^{-\rho} u_2$ is part of the plant equations, and there must obviously be another part of these equations that makes x increase. Assuming that x increases at a constant rate θ when the male is not creeping, the plant equation becomes:

$$\dot{x} = \theta - x^{-\rho} u_2 \qquad (9.12)$$

and therefore H is given by

$$H = e^{-\phi St} Su_1 x + Su_3 x - \lambda \theta + \lambda x^{-\rho} u_2 \qquad (9.13)$$

which is equation (9.5).

To find the behaviour that maximizes the objective function (equation (9.4)), H must be instantaneously maximized with respect to u. Inspection of equation (9.13) shows that there is a term $\lambda\theta$ which does not depend on u and so can be ignored. The optimal decisions can therefore be found from equation (9.11). Each of the three terms is compared, and the biggest one is chosen. It therefore follows that each u will take its maximum value when it is performed. This does not correspond to the behaviour of male newts, who reduce their display rate with each spermataphore deposited (Halliday, 1975; Halliday and Houston, 1978). Simulations of the model (reported in Houston, 1977) showed that the pattern of display, creep, spermatophore transfer could be obtained, but it was not easy to find values of the parameters that gave 'newt-like' behaviour. In particular, Houston was not able to obtain three successive spermatophore depositions by computer simulation. Given the large number of parameters involved, this could be no more than a result of bad luck in the choice of values, but combined with the failure to produce the dependence of display rate on S, it suggests that the model should be treated with scepticism. Nevertheless, it represents the type of exercise that is necessary if objective functions are to be obtained by the inverse optimality approach.

9.1.3 Discussion

The example of newt courtship can be used to illustrate a general procedure for attempting to construct optimization models from causal models. A deterministic optimization model requires that the model perform the activity that gives rise to the highest value of H at each particular point in time. If the causal model involves the assumption that the activity with the highest tendency is the one performed, then this tendency can be used as an indication of the terms in H during the relevant activity. In the case of NEWTSEX IV this strategy requires the assumption that the hope variable is related to H rather than to the costate variable alone, as Dempster assumed. In other words, during retreat display, a causal variable such as hope could be given by the portion of H that depends on retreat display, while the terms in H that depend on creep would give a curve below the curve of hope (see Fig. 9.1). The same relationship must also hold for the terms that depend on spermatophore transfer, but we will confine our attention to retreat display and creep. As well as simplifying the exposition, this serves as a reminder that, in the causal

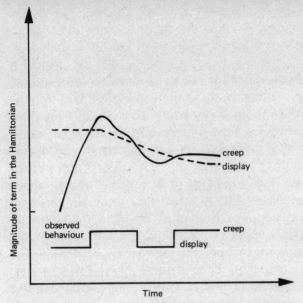

Fig. 9.1 Magnitude of the creep and display terms in H(t). The observed behaviour corresponds to the term that is largest. A causal model could be based upon competition among the terms in H.

model (NEWTSEX IV) spermatophore transfer occurs after a fixed duration of creep, and so the behaviour of hope is irrelevant once the transfer has started. Creep starts when the terms in H that depend on creep become greater than those that depend on retreat display. Because the consequences of creep are different from those of retreat display, both portions of H now decrease, with retreat display portion being less than the creep portion (see Fig. 9.1). If an attempt is made to construct an H that does not depend explicitly on time, then the fact that H is a constant along the optimal trajectory can be used to place limits on the terms of H that do not depend upon retreat display and creep.

The results of such a procedure would, of course, depend on the plant equations of the causal model. This dependence will generally hold, and must be borne in mind when considering the uniqueness of extremal principles as descriptions of behaviour. Houston (unpublished manuscript), gives the following example of the effect of modifying the plant equations. If it is required to move a deficit from an initial value x_0 to zero such that $\int_0^T u^2 \, dt$ is minimized, then if $\dot{x} = -ru$, the optimum solution involves keeping u constant. Modifying the plant equation to $\dot{x} = bx - ru$ results in the optimal u being a negative exponential if b is a negative constant. In other words, the optimal solution to this problem has the same form as the optimal solution to the quadratic problem considered in section 8.2.2.

From the above discussion it may seem that the inverse optimality approach is so fraught with difficulties as to be unworkable. While it is true that there may be difficulties in the interpretation and uniqueness of results, the exercise in itself can be worthwhile. For example, the work on newt courtship led to the suggestion that oxygen supply can increase by means of skin respiration during the relatively unenergetic creep portion of the courtship. This suggestion implies that the male deploys the creep behaviour as a means of prolonging the courtship and it can account for the fact that the male oscillates between creep and display more frequently in the later stages of courtship, when his oxygen supply is presumably lower (McFarland, 1977). Not only is this suggestion an interesting example of a physiological prediction made on a behavioural basis, but it is also open to empirical investigation. The idea that oxygen supply actually increases during creep behaviour has not yet been put to experimental test, but it is reinforced by the observation that this particular species of newt (*Triturus vulgaris*) not only has a more prolonged courtship than related species, but also has a larger crest which probably enhances skin respiration (Halliday, 1975).

Although the goal function suggested by an inverse optimality exercise cannot be proved to be unique, there are, nevertheless, ways in which its validity can be substantiated. In some cases it may be possible to determine how closely the objective function postulated (taken to be the goal function) resembles a cost function obtained from field research. Even where this is not possible it will sometimes be profitable to ascertain whether or not the proposed function makes ecological sense. In other cases it may be possible to establish the plausibility of an objective function from evolutionary principles. This is the approach advocated by Freeman and McFarland (1981).

9.2 Decision rules

Although animals may be built to perform optimally with respect to their goal function, this may not always be reflected in their behaviour. Such a disparity between the goal function and the function that appears to be maximized or minimized follows from the existence of relatively simple decision rules ('rules-of-thumb') which can produce the optimal behaviour, or a close approximation to it. In other words, it is not necessary for an animal to perform a cognitive evaluation of the goal function, using the optimality procedures we have discussed in Chapters 7 and 8. Instead, the animal can follow rules that do not require complicated calculations. These rules may, however, have a limited applicability. Under some circumstances they can result in suboptimal behaviour. Examples of the 'slavish' adherence to rules are well known in ethology. Thus Baerends (1941) investigated the ability of the digger wasp *Ammophila* to provision a number of nests simultaneously, bringing the correct number of caterpillars to each nest each day. Each nest is inspected once per day and this inspection determines the amount of provisioning for that day regardless of the state of the nest during the rest of the day. In nature this may be a near-optimal rule because nests normally remain undisturbed. When Baerends changed the contents of the nest he found that the wasps failed to respond appropriately once having made the inspection visit for the day.

We illustrate the rule of thumb idea with a simple example involving feeding behaviour. In Chapter 8 we discussed the problem of minimizing

$$\int_0^\infty a_1 x^2 + a_2 u^2 \, dt \qquad (9.14)$$

where x is the food deficit, u is the feeding rate, and $\dot{x} = -ru$. To simplify the notation, we assume that $a_1 = 1$ and denote a_2 by A^2. The cost function then has the form $C = x_1^2 + A^2 u^2$.

We will assume that A depends on the risk of the animal being taken by its predators. For a given environment, with a certain frequency of predators, the cost function will be given by a certain value of A. (We assume for the purposes of the argument that the correct form of the function is quadratic.) In environment 1, let the correct value of A be A_1. If the animal is well adapted to environment A, its goal function will be the same as the cost function and hence will be given by

$$G = x^2 + A_1^2 u^2$$

We know from Chapter 8 that the optimal feeding rate is given by

$$u(t) = (x_0/A_1) e^{-(rt)/A} \qquad (9.15)$$
$$= (x(t))/A_1 \qquad (9.16)$$

Equation 9.16 gives a simple rule for the optimal behaviour – the animal keeps its feeding rate proportional to its food deficit, the constant of proportionality being $1/A_1$. We now move the animal to environment 2, for which the appropriate value of A is A_2. If the animal generates its behaviour on the basis of equation (9.16), with A_1 a fixed property of its nervous system, then the animal's behaviour will not change. (The possibility of the parameter A not being fixed is discussed in the next chapter.) Because the animal still behaves in the same way, it will still minimize the same goal function, but this function will no longer be the cost function.

In environments 1 and 2 the plant equation is

$$\dot{x} = -ru,$$

but in environment 3 this relationship no longer holds. For the purposes of the argument, we consider the plant equations of the form

$$\dot{x} = d(x, u, t) \tag{9.17}$$

but assume that neither the animal nor the experimenter can establish the function d. All the animal does is follow the control law $u = x/A$, and the experimenter cannot observe the deficit, x, and so assumes that $\dot{x} = -ru$. Without any claims of plausibility, assume that

$$d(x, u, t) = \frac{-r(u+M)}{((x/A)+M)}$$

where r and M are constants. Following the rule $u = x/A$ means that

$$\dot{x} = -r\frac{((x/A)+M)}{((x/A)+M)}$$

$$\therefore \quad x(t) = x_0 - rt$$

and

$$u(t) = \frac{x_0 - rt}{A} \tag{9.18}$$

Equation (9.18) shows that as a result of the modified plant equation and the control law, the feeding behaviour decreases linearly with time. As a result, the behaviour no longer minimizes the quadratic goal function. That is not to say that the behaviour cannot be described in terms of the minimization of an objective function, but it does mean that the objective function would not be the same as the goal function (or the cost function).

Things will, of course, be a lot more complicated than this, especially when we consider rules that are only approximately optimal at the best of times. Some examples are given in Houston (1980).

Our general point is that if the animal does not detect a change in the consequences of its behaviour, it behaves as usual. For example, McFarland (1971) reported an experiment in which doves (*Streptopelia risoria*) were trained to work for small water rewards. After the initial training the doves were tested daily for alternating periods of 7 days using distilled water or 0·5% saline as a reward. This saline concentration is not discriminated from distilled water by doves, so differences in injection pattern must be due to the systemic physiological consequences of drinking (McFarland and McFarland, 1968). McFarland (1971) obtained evidence that the doves behaved as though their drinking water had not changed when they were tested on the first day after a switch from saline to distilled water, or vice-versa. On subsequent days, however, the doves behaved as if their short-term satiation rule had been recalibrated as a result of the systemic consequences of the previous day's consumption of the test solution. It is well known that alterations in food or water selection can result from delayed physiological consequences of injection (Rozin and Kalat, 1971; McFarland, 1973). McFarland (1971, p. 111) postulated that the optimal solution for a dove was to obtain critical damping in the drinking control system. Under such conditions ultimate satiation is achieved as quickly as possible, without causing systemic overhydration. His results were consistent with the view that deviations from critical damping led to a recalibration of the short-term satiation mechanisms. In other words, when the change in the environmental situation was not immediately detectable, the doves persisted with the rule-of-thumb that had previously produced the optimal outcome. If they subsequently discovered (through the ultimate physiological consequences of their behaviour) that the optimal consequences were no longer obtained, then they learned to alter the rule-of-thumb. McFarland maintained that this was an example of adaptive control in animals, and in Chapter 10 we discuss the relationship between learning and adaptive behaviour.

Our present point is that rules-of-thumb applied in a particular environment may give behaviour that is close to the optimal. If an environmental change is undetected by the animal, it will continue to apply the rule. If the plant equations remain unchanged, the animal's behaviour will be unchanged, but it will no longer be so adaptive. This might be the case where a predator or a poisonous food becomes part of the animal's environment without being detected. Such cases result in a change in the cost function, but no change in the goal function or in the rules-of-thumb deployed by the animal.

If the undetected environmental change does result in a change in the plant equations (as in the case of McFarland's dove experiment), then there will be a consequent change in the animal's behaviour even though the animal is using the same decision-rules (McFarland found that the drinking pattern changed in the predicted manner). If this happens, the animal may be able to learn to change its decision rules. We argue in Chapter 10, however, that the animal cannot learn to change its goal function. It might, however, be able to deploy evolutionarily preprogrammed changes in the goal function, and this possibility is also discussed in Chapter 10.

The difficulty for an investigator is to distinguish between the various ways that an animal can adapt to environmental change. We discuss these in the next section.

9.3 Adaptation to environmental change

The two most important factors in considering the adaptation of individual animals to changes in the environment are (1) the extent to which the animal is designed to cope with change, and (2) the ways in which the animal detects environmental changes.

Animals that have evolved in stable, unchanging environments are likely to make decisions on the basis of relatively stereotyped sets of rules tailored to suit the ecological niche. Animals that have evolved in situations where they are likely to be subject to environmental changes during their lifetime will have various means of adaptation, the most important of which are acclimatization and learning. For the sake of example, let us consider an animal that is able to adapt to a range of environments $1, 2, 3, \ldots$. When in environment 1, it could use cue x_1 to set the value of a parameter (see equation (9.14)), which is related to the risk of the animal being taken by its predators. The analogous cue in environment 2 is x_2, and the respective cost functions for environments 1 and 2 are as follows:

$$C_1(A) = x^2 + A_1^2 u^2$$
$$C_2(B) = x^2 + A_2^2 u^2$$
(9.19)

The animal uses the relevant habitat cues to change its decision rules, and this is equivalent to setting the value of A in the decision rule corresponding to part of its goal function. This is one possible solution to the problem of differential predation in different habitats. The animal does not have to detect the presence of predators directly (x_1 and x_2 can be any feature of the respective habitats), but is pre-wired to behave differently in the two environments, being cautious in one and bold in the other. An alternative solution to the problem is for the cues x_1 and x_2 to be directly related to the presence of predators, these cues then influence the animal's motivational state so that the appropriate cautious and bold behaviour results from a wariness factor in the state of causal factors (see section 8.2.2). There need then be no differences between the goal functions in the two environments, although the differences in cost function (equation (9.19)) would remain. A third possible solution is that the plant equations change as a result of acclimatization or learning. The goal function may remain unaltered, but the behaviour is different in the two environments as a result of the changed plant equations. Acclimatization differs from the previous two options in that it entails some cost in changing the animal's physiology.

The probable solutions to the problem of differential predation in two or more environments may now be summarized: (1) changes in the animal's decision rule cued by habitat factors; (2) changes in wariness directly cued by detection of predators; and (3) changes in the plant equations resulting from acclimatization or learning. (Other possibilities are not included here on grounds of their general implausibility. These include learned modification of the goal function, and cognitive appraisal of the different environments. Some of these issues are discussed in Chapter 10.)

The relative importance of these possible solutions is difficult to assess. Houston and McFarland (1980) address the problem and discuss the case of the female Burmese red jungle fowl (*Gallus gallus spadiceus*) which spends 99% of her time on the nest during incubation, has little time for feeding, and progressively loses weight (Sherry, Mrosovsky and Hogan, 1980). They conclude that acclimatory changes in digestive efficiency are unlikely on physiological grounds and note that such a change would be expected to result in changes in the demand function for food. They also conclude that changes in the goal function (specifically, in resilience; see section 7.1.2), such as probably occur in hibernating animals, are unlikely to occur during incubation because of evidence that food intake continues to be regulated during incubation. Houston and McFarland (1980) conclude that an acclimatory change in the set point of energy regulation is the most likely form of adaptation. They note that this is the only explanation that would not be expected to lead to changes in the elasticity of demand for food. The relevant demand experiments have not been carried out, but the discussion of this example illustrates that it is possible to arrive at some means of discriminating between the possible explanation that can be tested experimentally.

Some generalizations can be made about the three probable solutions to the problem of differential predation outlined above. In cases where the environments 1 and 2 are very different, as with many migratory birds, we might expect to see changes in the animal's goal function on the grounds that large changes in plant equations are difficult to engineer. If the goal function does not change when there are large changes in the cost function, then any adaptation to environmental change is likely to be approximate and suboptimal. The two other probable solutions to this problem are geared to an unchanging goal function, which is bound to be different to the relevant cost function. This may not matter much if the environments 1 and 2 are not very different, or if the animal experiences one environment for a short time only. However, these strategies have a counter-advantage in situations where there is a permanent change in the environments 1 and 2. Suppose, for example, the cost function in equation (9.19) were to change, viz:

$$\begin{aligned} C_1 &= x^2 + (A_1 - E_1)^2 u^2 \\ C_2 &= x^2 + (A_2 - E_2)^2 u^2 \end{aligned} \qquad (9.20)$$

Preprogrammed changes in the goal function of an animal moving from environment 1 to environment 2 would no longer be relevant, but strategies of adaptation involving wariness, acclimatization or learning could still be beneficial. If the permanent environmental changes involved factors which could only be represented in the goal function, then the animal could not learn about them (see Chapter 10). Under such conditions it might well be better for the animal to rely on direct detection of predators and the consequent changes in wariness.

To summarize, animals evolved in a stable environment may rely on genetically determined strategies of adaptation involving contingent changes in the goal function. Animals in less stable environments probably do better to rely on acclimatization and learning. Adaptation by learning involves further consideration of optimality principles which are discussed in the next chapter.

9.4 Summary

In this chapter we introduce the inverse optimality approach and illustrate it by a particular example. Although this approach is capable of providing an objective function that can account for the observed behaviour of the animal, problems of uniqueness remain. In particular, because there may be an alternative objective function, there is no guarantee that an apparently successful objective function is the same as the goal function embodied in the animal.

Animals may employ decision-rules which give rise to approximately optimal behaviour in the natural environment. Changes in the environment will mean that the behaviour of an animal with rigid decision rules will be suboptimal. Various possible means of adaptation to environmental change are discussed.

10 Learning and optimization

Throughout ontogeny, animals may acquire new decision-rules, and may be subject to changes in goal function or plant equations. Some such changes are genetically predetermined, but others are acquired through experience. Learning is a part of ontogeny that is contingent upon environmental events.

When an animal learns to respond to an environmental stimulus in a new way, its behavioural repertoire undergoes a permanent alteration. Learning is an irreversible process, and although an animal may extinguish or forget learned behaviour, the animal can never revert to its previous state. Most apparent unlearning is learned inhibition of previous learning (Mackintosh, 1974).

Any change in the animal's makeup is likely to alter its fitness. Learning is a process that is widespread in the animal kingdom and we would expect it to bring about beneficial changes on average, because the learning processes themselves have long been subject to natural selection. The idea that learning is adaptive has long been recognized by the more biologically minded students of behaviour (e.g., Thorndike, 1911; Tinbergen, 1951; Lorenz, 1965), but the force of the argument has only been relatively recently recognized by learning theorists (e.g. Rozin and Kalat, 1970; Seligman, 1970; Revusky and Garcia, 1970; Staddon, 1976).

In this chapter we distinguish between three aspects of learning: preprogrammed learning (section 10.1), learning in predictable situations (section 10.2), and learning in unpredictable situations (section 10.3). To appreciate these distinctions we have to consider the role of evolution in shaping the animal's learning capacities. Situations that are predictable from an evolutionary point of view, such as the onset of sexual maturity, are likely to lead to a preprogrammed form of learning such as sexual imprinting. However, a situation may be unpredictable from an evolutionary point of view but predictable by individual animals. Thus in a given situation an animal may be able to predict the progress that it will be able to make through learning to cope with the situation. We refer to this as learning in a predictable environment. In many instances the individual animal may not be able to predict the consequences of learning, but evolution may nevertheless have decreed that it should invest in learning in this type of situation. We call this learning in unpredictable situations.

10.1 Preprogrammed learning

Adaptation of animal behviour to an unvarying environment requires decision rules, but the rules may be a fixed property of the individual. Invariant features of the environment, such as the properties of gravity, are generally responded to in a stereotyped manner by all members of a species. Thus the anti-gravity and postural reflexes of pigeons are consistent and stereotyped (Mittelstaedt, 1964; Delius and Vollrath, 1973).

Cyclic or other predictable changes in the environment can also be handled on the basis of fixed decision-rules. As explained in Chapter 3, biological rhythms can be incorporated into our mathematical framework in a number of ways. From the point of view of optimality theory, circadian rhythms have their effects as changes in the plant equations, resulting either from cyclic endogenous (command) factors or from exogenous (cue) factors. Circannual or seasonal rhythms can also be represented as cyclical changes in the plant equations. In section 2.1 we saw that acclimatization always involves a change in physiological state, and can always be represented as an acclimatization vector in physiological space. McFarland (1978a) pointed out that the term acclimatization is a convenient label under which to group all slow-acting physiological adjustment, including such annual events as hibernation, migration and reproductive cycles. In the case of sexual behaviour, there would be a region of the physiological state space where sexual behaviour is discouraged by the attenuation of motivational command vectors, and another region where it is allowed free reign. Thus under certain conditions sexual behaviour is an expensive luxury in that it endangers physiological stability, consuming energy and taking up time that might profitably be devoted to other behaviour.

The competing claims of different aspects of the animal's behavioural repertoire are discussed in Chapter 7. For our present purposes it is sufficient to note that circadian, circalunar and circannual rhythms of behaviour can be accounted for in terms of preprogrammed changes in the plant equations. Such cyclical events are predictable from an evolutionary point of view and in many animals they are programmed on the basis of endogenous clocks (Cloudsley-Thompson, 1980; Pengelley, 1974).

Changes throughout the animal's life history may also be predictable from an evolutionary point of view. In section 8.3 we saw that considerations of life history strategies can lead to the conclusion that an animal's goal function may change throughout its life. Where such changes are genetically preprogrammed the problem can be tackled by means of conventional optimality theory (Freeman and McFarland, 1981). In other cases, however, the goal function may change as a result of preprogrammed learning, enabling the animal to adapt at an early stage of life to features of its habitat or social position that may be different from those it has experienced by its parents. For example, Bateson (1979) argues that some young birds need to discriminate between the parent that cares for it and other members of its species, because parents themselves discriminate between their own and other young and may attack young that are not their own. This type of selective pressure leads to the evolution of filial imprinting, involving specific periods of the life history during which the young animal learns to recognize its parent(s). Bateson (1979) also argues that sexual imprinting enables an animal to learn the characteristics of its close kin so that it can subsequently choose a mate that is slightly different, thus attaining an optimal balance between inbreeding and outbreeding. The sensitive period for sexual imprinting is usually later than that for filial imprinting, probably because the appearance of siblings changes rapidly early in life and a more reliable assessment of their individual characteristics can be made later in development (Bateson, 1979).

Our point is that when the problems facing the juvenile animal differ slightly from generation to generation the optimal life-history strategy should include processes that are contingent upon the situation at particular stages of the life cycle. The various types of imprinting provide examples of this type of phenomenon.

Imprinting is evolutionarily preprogrammed in the sense that it does not require feedback of information about the value of alternative courses of action. The imprinting animal learns willy nilly the features of the parent, sibling or habitat presented to it. This type of learning may well result in permanent changes in the animal's goal function, in contradistinction to other learning situations (see below) where we claim that it is impossible for the goal function to change. In the case of preprogrammed learning, fitness is increased when the goal function becomes more like the cost function, and it is inherent in the imprinting situation that the goal function will change in this direction and not in any other direction. For example, in sexual imprinting we can imagine that certain species characteristics form a gradient $x_a, x_b, \ldots$, which might represent plumage colouration, for instance. If a young animal's siblings have the colour range $x_d - x_g$, then the optimal colouration for a sexual partner might be $x_j - x_n$. The cost function will therefore include an $x_j - x_n$ term. By learning that the colouration of its siblings is $x_d - x_g$ our juvenile will be predisposed to prefer sexual partners with the $x_j - x_n$ colouration (according to Bateson's views). In other words, the imprinting process automatically results in conformity between the goal and cost functions for sexual behaviour in later life.

10.2 Learning in a predictable environment

If the environment changes in some crucial respect, reliance upon fixed decision rules will be maladaptive. It will sometimes be the case that such changes are completely unpredictable, but it may also be the case that the environmental changes, though unpredictable from an evolutionary viewpoint perspective, are predictable by the individual animal. Thus an individual animal seeing conspecifics feeding in a new location may join them without having learned the technique appropriate to the particular food source. The individual can, however, predict that it will be able to learn the techniques within a certain period of time.

The learning process has long been subject to natural selection so that, whatever changes occur, they must be in a direction that tends to improve the animal's fitness, and not be merely random or experimental changes. In terms of optimality theory the changes induced by learning must lead to an improved trajectory in the state space. What parameters, in a Hamiltonian type of formulation, could be changed in such a way to lead to such improvement?

From section 8.2, $H = \lambda^*$ (plant equations) – goal function.

We can envisage changes in either of these two parts. However, McFarland (1978b) argues that changes in the goal function are not possible in principle and that for such changes to constitute an improvement, the goal function (McFarland, 1978b, uses the term objective function) would have to be altered in a manner which made it more like cost function, because the cost function represents the best that can be attained in a given environment.

'To be adaptive in the biological sense, such changes in the objective function would have to bring the cost and objective functions closer. Learned modification of the objective function cannot, of course, be based upon information about the nature of the cost function, unless we are prepared to allow that the individual understands the function of its own behaviour.' (McFarland, 1978b, p. 75.)

Learning about unpredictable environmental changes must be based upon some kind of feedback, or knowledge of results. The feedback must be evaluated in terms of criteria that are a property of the individual animal. These criteria are themselves incorporated into the goal function. The animal cannot learn on the basis of any other criteria, so it follows that the goal function cannot itself be changed by learning. The only type of learning that could lead to changes in the goal function is preprogrammed learning that is anticipating (in the evolutionary sense) predictable changes in the environment. As we point out in section 10.1, the juvenile is predisposed to learn certain types of things, such as the nature of its habitat, of its parents, and of food sources. This type of learning is a form of contingent maturation. It is also conceivable (see McFarland, 1978b) that individuals can learn to coordinate with others in modifying the environment in such a way that the cost function is brought closer to the goal function. In our view, modification of the goal function is not a possible way of learning to cope with evolutionary unpredictable changes in the environment.

If the parameter changes induced by learning do not occur in the goal function, they must occur in the plant equations. There are two possible candidates, r and k. The resource availability is defined as $r = -\dot{x}/u$. The constraint parameter k is defined as the ceiling on u due to the limiting behavioural abilities of the animal.

Changes in r need not only be related to one activity but might also be seen as changes in activity u_1, that result in consequences for another activity u_2. For example, an animal may visit a new location (change in u_1) and discover that the consequences of eating a particular food (u_2) are different from previously. Since the consequences of u_2 are $\dot{x} = r_2 u_2$, it is r_2 that has changed. The consequences of u_2 may relate to the nutritive properties of the food, to the danger of eating in the new location, etc. These are all aspects of r_2 that can provide a basis for learning. Evaluation of the new consequences of u_2 (in terms of the goal function) leads to learning to exploit or to avoid the new location.

Alternatively we can imagine that an animal may accidentally eat a novel food which has memorable consequences. Many animals quickly develop a learned aversion to novel foods that have deleterious post-ingestive consequences (Revusky and Garcia 1970; Rozin and Kalat, 1971; McFarland, 1973). Animals may also learn to ignore irrelevant uniformative stimuli (Mackintosh, 1973). Thus, subtle changes in activity u_3, such as attentional changes, may have particular consequences. Such changes will appear as changes in resource availability r_3.

Changes in k result in alteration of the constraints on behaviour rate u. This possibility is discussed by McFarland (1978b). An animal could alter the constraints upon its behaviour by learning to handle situations with greater skill, or by acquiring new behaviour. For example, an animal that could learn to reduce the handling time of food items could feed at a higher rate. An animal that could learn to reduce the degree of incompatability between alternative activities could effectively reduce the constraints upon its choices (see McFarland, 1978b). Thus a fox may learn to mark its territory en route to a possible feeding site. The differences between learning to alter r and k may seem subtle, but they can be shown to have markedly different effects in terms of optimization.

At this point it may be helpful to reconsider the simple example with a unidimensional state x and a single behaviour u

$$\dot{x} = -ru$$

and a quadratic goal function (see Chapter 8)

$$G = x^2 + u^2.$$

The optimal solution is given by

$$x(t) = x_0 e^{-rt}$$

This exponential type of behaviour (see equation (8.11)) is typical of an unconstrained situation. However, there may be a ceiling on the rate at which behaviour can be deployed. If x is less than k (see Fig. 10.1), the optimal behaviour is not affected by the constraint, but when $x > k$ the decremental cost is minimized when u is as close as possible to the point where $u = x$, i.e. $u = k$.

There is now sufficient information to calculate the total costs involved in the trajectory resulting from any value of k (Tovish, 1981). Thus in terms of Fig. 10.1, the best route is marked by the thick line.

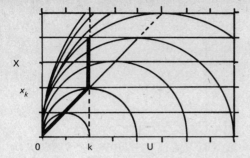

Fig. 10.1 A simple unidimensional example in which rate of behaviour u is plotted against state x. The semicircles are isoclines of incremental cost. k is the limit on u, so that $u = k$ when $x > x_k$, and $u = x$ when $x < x_k$. An optimal trajectory is shown by the thick line.

10.2.1 Choice between environments

Suppose an animal has been familiarized with and recognizes several environments that have characteristic values of r and k. The environments offer different opportunities for changing x in accordance with the equation $\dot{x} = -ru$ and differ in the extent to which behaviour is constrained. For example, food may be plentiful in one environment (high r) and scarce in another (low r). It may be relatively accessible in one environment (high k) and inaccessible in another (low k). For any given state we assume that the animal can make an accurate evaluation of the advantages and disadvantages of each environment. Thus when $x_0 = 1$ the possible alternative environments can be thought of as points in the (r, k) space illustrated in Fig. 10.2. In this figure the isoclines connect environments in which it would be equally costly to change the state from $x = 1$ to $x = 0$. Thus in a choice between two points (e.g. points A and B) that lie on different isoclines, the point on the upper isocline (i.e. point A) is to be preferred, because it would cost less to arrive at $x = 0$ in this environment.

Improvement in resource availability r or accessibility k lowers the cost of attaining $x = 0$. However, improvement of k beyond $k = 1$ has no effect on the cost, so for a given r there is a minimum cost that can be achieved. This is not the case with k which can always be compensated by a sufficiently high r.

Fig. 10.2 applies only to the state $x = 1$. If there were a third axis for x, we would find that environments with equal associated costs when $x = 1$ would no longer be the same. At smaller initial values of

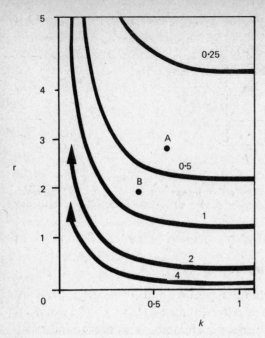

Fig. 10.2 Resource availability r versus resource accessibility k. The isoclines connect environments in which it would be equally costly to change the state from $x = 1$ to $x = 0$. In a choice between two environments (e.g. A and B), the one on the upper isocline (i.e. A) is to be preferred, because it would cost less to arrive at $x = 0$ in this environment.

x the environment in which food is more abundant would yield a lower-cost solution, as illustrated in Fig. 10.4, p. 179.

10.2.2 Learning to change the parameters

Learning should be evaluated in terms of its entire period of usefulness. Strictly, this would require a calculation covering a large portion of the animal's life. For our present purposes we can assume that whatever is learnt becomes quickly outdated because the environment changes. Each test situation thus requires a new learning effort.

Different animals have different learning abilities and we have to make some assumptions about this. It is convenient, though not essential, to assume the traditional learning curve (Kimble, 1961)

$$k = k_0(1 - e^{-q\tau}) \tag{10.1}$$

where k = constraint parameter, k_0 = asymptotic level of constraint, q = learning ability, τ = experience.

This formulation assumes that learning serves to change the constraint k which is preventing the animal from exploiting a situation as fast as it might otherwise. Initially k is low and the animal has little idea how to exploit the situation. We envisage that it learns to do this as a function of experience τ which is the time actually spent engaged in the behaviour u:

$$\frac{d\tau}{dt} = \frac{u}{k} \qquad (10.2)$$

When the animal spends all its time in u, then $u = k$ and it accumulates experience as quickly as time passes, but when $u < k$ experience accumulates more slowly.

We can illustrate this phenomenon by means of an example in which $r = q$ and $x_0 = k_0$. The animal starts with zero knowledge of the situation and its behaviour u has the effect of changing both k and x. Figure 10.3 shows the behaviour of the three variables u, k and x under three sets of conditions:

(1) There is a cost associated with the state x but none with the behaviour u. The animal acts as fast as it can until x is reduced to zero, which occurs at time rt_x (see thin lines in the figure). Under these conditions the animal is learning as fast as possible, $\tau \equiv t$ and k follows the characteristic learning curve (thin line Fig. 10.3 I). There may remain some room for improvement (dashed thin line) beyond the satiation point. This could come into play with larger initial values of x.

(2) There are costs associated with both x and u. The behaviour is initially the same as in the previous case (thick line Fig. 10.3 III). The animal learns as fast as possible so long as inexperience dictates that $u < x$. At time rt_1, when $u = x$, the cost of the behaviour begins to have an effect similar to that discussed in section 8.2.2, and there is a typical exponential decline in u. This results in a deceleration of k as not all time is devoted to learning (thick line Fig. 10.3 I). It also means that x declines more slowly (thick line, Fig. 10.3 II).

(3) The animal behaves at a constant rate k_f, no learning being involved (thick dashed line, Fig. 10.3 I and III). There is a linear decrease in x up to time rt_f where $u = k$ (Fig. 10.3 II). The state x thereafter declines exponentially as expected on the basis of equation 8.11. To facilitate exposition, k_f has been chosen such that the integrated costs are the same as those of the variable (learned) constraint.

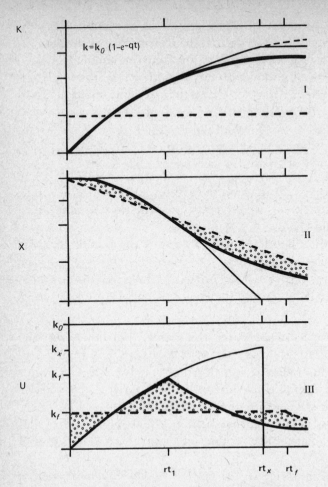

Fig. 10.3 The behaviour of k, x and u in a simple example in which $r = q$ and $x_0 = k_0$. When there is a cost associated with state x, but none with behaviour u, the animal changes its state as fast as it can (thin lines). When there are costs associated with both x and u the animal learns as fast as possible up to the point (rt_1) where $u = x$ after which there is a typical exponential decline in u (thick lines). The dotted line shows the situation where no learning is involved and u is constrained at k_f. (After Tovish, 1981.)

Initially, the low rate of the as yet unlearned behaviour incurs a lower cost than the fixed rate k_f, but as learning progresses, higher costs are incurred. These differences are indicated by the shaded areas in Fig. 10.3 III. (For a true comparison the values should be squared.) Similarly, the initial low rate of unlearned behaviour leaves the state x relatively unaffected so that higher costs are initially incurred. As learning progresses there is a higher rate of change of state and the costs are lower than in the unlearned case (see shaded areas in Fig. 10.3 II).

These comparisons show that learning provides an initially costly strategy which becomes less costly as a function of experience. If k_f were larger than that shown in Fig. 10.3 III, the non-learning strategy would incur less overall cost than the learning strategy, and *vice versa* if it were smaller. If learning could get off to a quicker start ($q > \sigma r$) the learning strategy would incur less cost, and *vice versa* when $q < r$. Thus certain combinations of k_f and q/r would result in isoclines of equal cost on a plot of k_f/k_0 versus q/r, such as that illustrated in Fig. 10.4.

We can imagine an animal presented with a choice between two environments: one unexplored but reliably cued as being learnable at rate q_1 and another less profitable but already mastered. This choice can be specified as a point on the plot shown in Fig. 10.4. We can determine which environment the animal should choose when in a

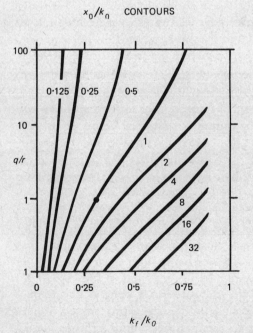

Fig. 10.4 Isoclines of x_0/k_0 in a space of learning ability (q/r) versus degree of existing mastery (k_f/k_0). The situation illustrated in Fig. 10.3 is indicated by the point at $q/r = 1$ and $x_0/k_0 = 1$. The coordinate of this point along the abscissa is the k_f/k_0 value obtaining in Fig. 10.3. (After Tovish, 1981.)

given state x. If the point lies to the right of the appropriate isocline the animal should choose the mastered environment and *vice versa*.

Learning is a strategy that can be deployed according to the circumstances. To be worthwhile, the exploitation of a learning situation must depend upon a number of considerations. These include the learning ability of the animal (q), the extent to which the situation can be exploited by learning (k_0) and the apparent utility of alternative courses of action that do not involve learning (k_f). Our simple example serves to illustrate these points.

10.3 Learning in an unpredictable environment

So far we have been considering completely deterministic problems. The solution to such problems can always be written as a function of time (see Houston, 1980), which emphasizes the fact the animal is not gaining any information. Instead, its performance changes in an exactly specified way, so that the future consequence of any decision is known in advance. In this section we describe a framework for cases in which the parameters of the problem are not known in advance, so that the animal's task is complicated by the need to gain information. A more detailed account is given by McNamara and Houston (1980).

As an illustrative example, consider an animal faced with two sources of food. Each source, which we will call a patch, has a certain probability of yielding a food item if chosen. These probabilities will be denoted by p_1 and p_2 corresponding to patches 1 and 2, respectively. To eliminate any complications arising from the cost of changing from one patch to another, we assume that the animal must return to a point between the patches before another response can be made. The animal's problem, loosely speaking, is to get as many food items as possible. It is obvious that if p_1 and p_2 are fixed and known to the animal, it should always choose the patch with the bigger probability. When p_1 and p_2 are not known, the problem is known as a two-arm bandit problem and is much more difficult to solve. If nothing at all is known about the possible values of p_1 and p_2, then the problem is not well defined and no progress can be made. Some idea of the likelihood of various values of the probabilities is essential to the problem. This information is given by what is known as the prior distribution. This distribution represents the initial probability of various possible values of a parameter. For example, if it is known that p_1 is either 1/3 or 2/3 and each alternative is equally likely, then the prior probability distribution is

$$P(p_1 = 1/3) = 1/2$$
$$P(p_1 = 2/3) = 1/2$$

where $P(A)$ is the probability of the state A.

When the animal makes a response, it may obtain a food reward, but it also gains information which can be used to modify the prior distribution to give a posterior distribution of the likelihood of the various possible values given the outcome of the response. The posterior distribution is obtained from the prior using Bayes' rule, which is why the approach is often referred to as Bayesian. To describe this rule we require the concept of conditional probability.

The conditional probability of an event A under the hypothesis H, written $P\{A/H\}$ is given by the following formula:

$$P\{A/H\} = \frac{P\{AH\}}{P\{H\}}$$

where $P\{AH\}$ is the probability of both A and H.

We have phrased the conditional probability in terms of hypotheses, because the animal can be thought of as having a set of hypotheses about the value of the unknown parameter (as McNamara and Houston (1980), emphasize, there is no need to assume that any of the mathematics is performed consciously. We return to this point later). Let these hypotheses be $H_1, H_2, \ldots, H_n$. It is necessary to find the new probability of each hypothesis, given the information (event A) obtained when a response was made. In other words, we wish to find $P\{H_i/A\}$ for $i = 1, 2, \ldots, n$. From the definition of conditional probability.

$$P\{H_i/A\} = \frac{P\{AH_i\}}{P\{A\}} \tag{10.3}$$

and

$$P\{AH_i\} = P\{A/H_i\} \cdot P\{H_i\} \tag{10.4}$$

Furthermore,

$$P\{A\} = \sum_j \{A/H_j\} \cdot P\{H_j\} \tag{10.5}$$

Inserting equations (10.4) and (10.5) into equation (10.3) gives

$$P\{H_i/A\} = \frac{P\{H_i\}P\{A/H_i\}}{\sum_j P\{A/H_j\}P\{H_j\}}$$

which is Bayes' rule.

Note that $P\{H_i\}$ is the prior probability of hypothesis i, so the posterior is proportional to the prior.

The proportionality term

$$\frac{P\{A/H_i\}}{\sum_j P\{A/H_j\}P\{H_j\}} = \frac{P\{A/H_i\}}{P\{A\}}$$

is often known as the likelihood ratio.

To show how Bayes' rule works, we return to the example of the prior for p_1 (page 181).

Let

H_1 be ($p_1 = 1/3$)

H_2 be ($p_2 = 2/3$)

and we assumed

$P(H_1) = P(H_2) = 1/2$.

Now imagine that the event A is making a response in patch 1 that results in a reward. Then

$P\{A/H_1\} = 1/3$

$P\{A/H_2\} = 2/3$

and

$$\sum_j P\{A/H_j\}P\{H_j\} = (1/3 \cdot 1/2) + (2/3 \cdot 1/2) = 1/2$$

$$\therefore \quad P\{H_1/A\} = \frac{1/2 \cdot 1/3}{1/2} = 1/3$$

and

$$P\{H_2/A\} = \frac{1/2 \cdot 2/3}{1/2} = 2/3$$

These new probabilities of 1/3 and 2/3 contain all the current information about patch 1. The original prior probabilities can be forgotten, and the posterior probabilities can be used as the new prior, which will in turn be modified by the next observation.

It will be recalled that the two-arm bandit problem involves uncertainty about two parameters p_1 and p_2. One form of this problem considers p_1 and p_2 to have a prior probability which is beta distributed with parameters α and β (see DeGroot, 1970). This is very convenient mathematically, because the posterior will also be a beta distribution. The mean value of the prior is $(\alpha+1)/(\alpha+\beta+2)$ and if n responses are made to a patch and r food items are obtained, the resulting posterior has a mean given by $(\alpha+r+1)/(\alpha+\beta+n+2)$.

One way to formulate the problem of allocating responses to the patches is to consider the animal to have a total number N of responses and to be trying to maximize the expected number of items obtained. Such problems are usually solved by working backwards from the case when only one response is left, using the principle of dynamic programming (see Chapter 8). This is the approach used by Krebs, Kacelnik and Taylor (1978), although for computational reasons they constrain the behaviour to be of the form: alternate between the patches for $2M$ trials, then spend the remaining $N-2M$ trials in the better patch. The optimal policy does not have this form, however. No definite decision to 'exploit' one patch is made; the decision of which patch to choose depends on the posteriors and how many responses are left. It is instructive to compare this case with the one-arm bandit problem that arises when one of the patches has a known probability of giving a food reward. Because responses in this known patch do not provide any information, once it is optimal to make a response in this patch, all future responses will be made there. Thus if the first response should be made on the known arm, no responses will ever be made on the unknown arm. If this is not the case and the first response is to be made on the unknown arm, then, depending on the posterior and the remaining number of responses, the animal should remain on the unknown arm or switch to the known patch and remain there. In this case it can therefore be said that there is a decision to exploit the known patch, but it is not the case that there is a decision to exploit the unknown arm. The optimal policy requires that the animal only make the next response in the unknown patch if the posterior exceeds a certain value that depends on how may responses remain. As a general rule, when few responses remain the known patch is favoured unless the unknown patch is clearly better, but when many responses remain the unknown patch may be chosen even if its posterior is below the probability of the known patch. This is because information could show the unknown patch to be better. The gaining of information becomes less important as the number of remaining responses decreases.

Despite this caveat, it is likely that the model used by Krebs *et al.* captures the qualitative features of the allocation of responses required by the optimal solution. This is suggested by the fact that it performs almost as well as is possible, i.e. as an animal that makes all its responses to the better patch (Kacelnik, personal communication). Furthermore, Krebs *et al.* found that great tits faced with the two-arm bandit problem do not just choose the patch which appears to be better at the time. In other words, the birds do not immediately maximize the probability of getting a food item, but adopt a long-term approach which involves sacrificing current gain in order to obtain information which may increase future gain. As we said earlier, it is not necessary to assume that the animal does this by the mathematical procedures we have been describing. It seems more plausible to suggest that the animal uses simple rules-of-thumb that approximate the optimal policy. Some rules for the two-arm bandit problem are considered by Kacelnik (1979).

An important consequence of an animal using rules-of-thumb is that the rules may not be appropriate for the problem that the experimenter has decided to investigate. In particular, the bandit problems we have been discussing assume that the reward probabilities are constant but unknown, but as McNamara and Houston (1980) argue, it is likely that many animals are adapted to food patches that vary in quality, perhaps as a result of depletion by the foraging animal. As a result, a run of unrewarded responses may be interpreted as an indication that the patch's food supply has been exhausted. McNamara and Houston develop a model of this sort of problem and find how many unrewarded responses an animal should put up with before abandoning the patch. The model assumes that the animal knows the initial probability P that a response in the patch will be rewarded. The exact form of the solution depends on the prior distribution for the duration of food in the patch, the cost of making a response and the alternative sources of food, but in general the animal maximizes total expected gains minus costs by continuing for longer in the face of unrewarded responses when P is low than when it is high. This corresponds to the well-known resistance to extinction shown by partially rewarded animals (see Mackintosh, 1974, for a review).

10.4 Conclusion and summary

The study of learning as an optimizing process is in its infancy, and the proposals put forward in this chapter can only be regarded as tentative. We hope, however, that we have made a start in unravelling a complex subject, and that we have stimulated interest in a previously neglected area. Our conclusions may be summarized as follows.

An animal's learning capabilities are shaped by natural selection and may be evolutionarily preprogrammed or may be based upon feedback (reinforcement) from the consequences of behaviour. Preprogrammed learning may be based upon ontogenetic changes in the animal's goal function.

Learning which depends upon explicit feedback cannot be based upon changes in the goal function, and is therefore most likely to involve changes in those parameters which make resources more available (r_i) or more accessible (k_i).

In situations in which the animal is able to predict its future learning progress, it effectively has a choice of learning to exploit a new situation or of remaining with a known and mastered situation. The animal may also have a choice in the extent to which learning should be directed at modifications in r or in k. In some circumstances the optimal choices can be specified.

In situations in which the animal is not able to predict its future learning progress, its task is complicated by the need to gain information. The optimal solution to this type of problem generally involves sampling as well as exploitation of resources.

References and author index

The numbers in the square brackets refer to the page or pages in the text where mention of a given work or person is made.

Aczel, J. (1966). "Lectures on Functional Equations and their Application". Academic Press, N.Y. [65]
Adolph, E. F. (1972). Some general concepts of physiological adaptations. *In* Yousef, M. K., Horvath, S. M. and Bullard, R. W. (eds), "Physiological Adaptations", pp. 1–7. Academic Press, New York and London. [22]
Allison, T. and Cicchetti, D. (1976). Sleep in mammals: ecological and constitutional correlates. *Science, N.Y.* **194**, 732–734. [46]
Anderson, N. H. (1962). On the quantification of Miller's conflict theory. *Psychol. Rev.* **69**, 400–414. [60]
Anderson, N. H. (1970). Functional measurement and psychophysical judgement. *Psychol. Rev.* **77**, 153–170. [60]
Anderson, N. H. (1978). Methods and designs. Measurement of motivation and incentive. *Behav. Res. Methods & Instrumentation* **10**, 360–375. [58–60]
Arbib, M. A. (1966). Automata theory and control theory – a rapprochement. *Automatica* **3**, 161–189. [86]
Arbib, M. A. (1969). Automata theory. In Kalman, R. E., Falb, P. L. and Arbib, M. A. "Topics in Mathematical System Theory", pp. 163–233. McGraw-Hill, N.Y. [86, 87]
Arbib, M. A and Zeiger, H. P. (1969). On the relevance of abstract algebra to control theory. *Automatica* **5**, 589–606. [85, 87]
Aschoff, J. (ed.) (1965). *Circadian Clocks*. North–Holland, Amsterdam. [44]
Athans, M. and Falb, P. L. (1966). *Optimal Control*. McGraw-Hill, New York. [129]

Baerends, G. P. (1941). Fortpflanzungsuerhalten und Orientierung der Grabwespen, *Ammophila campestris. Tijdschr. Ent.* **84**, 68–275. [161]
Baerends, G. P., Brouwer, R. and Waterbolk, H. Tj. (1955). Ethological studies of *Lebisites reticulatus* (Peters). 1. An analysis of the male courtship pattern. *Behaviour* **8**, 249–334. [13, 49]

Baerends, G. P. and Kruijt, J. P. (1973). Stimulus selection. *In* Hinde, R. A. and Stevenson-Hinde, J. "Constraints on Learning", pp. 23–50. Academic Press, London. [39–41, 106]

Baldwin, B. A. and Ingram, D. L. (1967). Behavioural thermoregulation in pigs. *Physiology and Behaviour* **2,** 15–22. [37]

Bateson, P. (1979). How do sensitive periods arise and what are they for? *Anim. Behav.* **27,** 470–486. [171]

Baum, W. M. (1974). On two types of deviation from the matching law: Bias and undermatching. *J. exp. Analysis Behav.* **22,** 231–242. [61]

Bellman, R. E. (1957). *Dynamic Programming.* Princeton University Press. [130]

Benzinger, T. H. (1969). Heat regulation: homeostasis of central temperature in man. *Physiol. Rev.* **49,** 671–759. [37]

Berthold, P. (1974). Circannual rhythms in birds with different migratory habits. *In* Pengelley, E. T. (ed), "Circannual Clocks". Academic Press, N.Y. [44]

Bertram, B. C. R. (1978). Living in groups: Predators and prey. *In* Krebs, J. R. and Davies, N. B. (eds), "Behavioural Ecology", pp. 64–96. Blackwell Scientific Publications, Oxford. [138]

Bolles, R. C. (1967). "Theory of Motivation". Harper and Row, New York. [3]

Bolles, R. C. (1975). "Theory of Motivation", 2nd edn. Harper and Row, New York. [3, 70]

Booth, D. A. (1972). Conditioned satiety in the rat. *J. comp. Physiol. Psychol.* **81,** 457–471. [69]

Booth, D. A. (1977). Satiety and appetite are conditioned reactions. *Psychosomatic Medicine* **39,** 76–81. [69]

Booth, D. A. (ed.) (1978). "Hunger Models". Academic Press, London. [69, 70, 73, 74]

Born, M. (1939). Cause, purpose and economy in natural laws (minimum principles in physics). *Proceedings of the Royal Institution* **30,** 596–628. [154]

Bray, G. A., Booth, D. A., Campfield, L. A., Mogenson, G. J. and Stunkard, A. J. (1978). Hunger modelling: a discussion of the state of the art. *In* Booth, D. A. (ed), "Hunger Models", pp. 451–470. Academic Press, London. [70]

Budgell, P. (1970). Modulation of drinking by ambient temperature changes. *Anim. Behav.* **18,** 753–757. [37]

Budgell, P. (1971). Behavioural thermoregulation in the Barbary dove (*Streptopelia risoria*). *Anim. Behav.* **19,** 524–531. [37]

Cannon, W. B. (1932). "The Wisdom of the Body". Kegan Paul, London. [73]

Carlisle, H. J. (1969). Effect of fixed-ratio thermal reinforcement on thermoregulatory behaviour. *Physiol. Behav.* **4,** 23–38. [122]
Carpenter, F. L. and MacMillen, R. E. (1976). Threshold model of feeding territoriality and test with a Hawaiian Honeycreeper. *Science, N.Y.* **194,** 639–642. [94]
Charnov, E. L. (1976a). Optimal foraging: Attack strategy of a mantid. *Am. Nat.* **110,** 141–151. [95]
Charnov, E. L. (1976b). Optimal foraging: the marginal value theorem. *Theoret. Pop. Biol.* **9,** 129–136. [95]
Cloudsley-Thompson, J. L. (1980). "Biological Clocks". Weidenfeld & Nicolson, London. [171]
Clynes, M. (1961). Unidirectional rate sensitivity. *Ann. N.Y. Acad. Sci.* **93,** 946–969. [86]
Collier, G., Hirsch, E. and Hamlin, P. (1972). The ecological determinants of reinforcement in the rat. *Physiol. Behav.* **9,** 705–716. [122]
Cowie, R. J., Houston, A. I., Kacelnik, A., Krebs, J. R. and Tarpy, R. (unpublished manuscript). The effect of satiation and risk of predation on feeding behaviour in the great tit (*Parus major*). [140]
Craig, R. B., DeAngelis, D. L. and Dixon, K. R. (1979). Long- and short-term dynamic optimization models with application to the feeding strategy of the loggerhead shrike. *Am. Nat.* **113,** 31–51. [141]
Cunningham, J. P. and Shepard, R. N. (1974). Monotone mapping of similarities into a general metric space. *J. math. Psychol.* **11,** 335–363. [63]
Curio, E. (1975). The functional organization of anti-predator behaviour in the pied flycatcher: A study of avian visual perception. *Anim. Behav.* **23,** 1–115. [42]

Davies, N. B. (1977). Prey selection and the search strategy of the spotted flycatcher (*Musicicapa striata*), a field study on optimal foraging. *Anim. Behav.* **25,** 1016–1033. [94]
Davies, N. B. (1978). Ecological questions about territorial behaviour. *In* Krebs, J. R. and Davies, N. B. (eds), "Behavioural Ecology", pp. 317–350. Blackwell Scientific Publications, Oxford. [94]
Davies, N. B. and Houston, A. I. (1981). Owners and satellites: The economics of territory defence in the pied wagtail *Motacilla alba. J. Anim. Ecol.* **50,** 157–180. [94]
Davis, W. J., Mpitsos, E. J. and Pineo, J. M. (1974). The behavioural hierarchy of the mollusk *Pleurobranchaea* I. The dominant position of feeding behaviour. *J. comp. Physiol.* **90,** 207–224. II. Hormonal suppression of feeding associated with egg laying. Ibid., 225–243. [106]
Dawkins, R. (1976). "The Selfish Gene". Oxford University Press. [72]
Dawkins, M. and Dawkins, R. (1974). Some descriptive and explanatory

stochastic models of decision-making. *In* McFarland, D. J. (ed.), "Motivational Control Systems Analysis", pp. 119–168. Academic Press, London. [72]

Dawkins, R. and Carlisle, T. R. (1976). Parental investment, male desertion and a fallacy. *Nature, Lond.* **262,** 131–132. [131]

De Benedictis, P. A. Gill, F. A., Hainsworth, F. R., Pyke, G. H. and Wolf, L. L. (1978). Optimal meal size in hummingbirds. *Am. Nat.* **112,** 301–316. [94]

DeGroot, M. H. (1970). "Optimal Statistical Decisions". McGraw-Hill, New York. [183]

Delius, J. T. and Vollrath, F. W. (1973). Rotation compensating reflexes independent of the labyrinth: Neurosensory correlates in pigeons. *J. comp. Physiol.* **83,** 123–134. [170]

de Villiers, P. A. (1977). Choice in concurrent schedules and quantitative formulations of the law of effect. *In* Honig, W. K. and Staddon, J. E. R. (eds), "Handbook of Operant Behaviour", pp. 233–287. Prentice Hall, Englewood Cliffs, N.J. [125]

Dixit, A. K. (1976). "Optimization in Economic Theory". Oxford University Press. [117]

Doucet, P. E. and van Straalen, N. M. (1980). Analysis of hunger from feeding rate observation. *Anim. Behav.* **28,** 913–921. [72]

Eberhard, M. (1969). The social biology of polistine wasps. *Misc. Publ., Mus. Zool.* **140,** 1–101. University of Michigan. [145]

Edmonds, S. C. and Adler, N. T. (1977). The multiplicity of biological oscillators in the control of circadian running activity in the rat. *Physiology and Behaviour* **18,** 921–930. [47]

Elgerd, O. (1967). "Control Systems Theory". McGraw-Hill, N.Y. [85]

Emlen, J. M. (1966). The role of time and energy in food preference, *Am. Nat.* **100,** 611–617. [95]

Emlen, S. T. (1974). Migration: orientation and navigation. *In* Farner, D. S. and King, J. R. (eds), "Avian Biology", **V,** 129–219, Academic Press, N.Y. [47]

Epstein, A. N., Fitzsimons, J. T. and Rolls, B. J. (1970). Drinking induced by injection of angiotensin into the brain of the rat. *J. Physiol., Lond.* **210,** 457–474. [5]

Farner, D. S. and Lewis, R. A. (1971). Photoperiodism and reproductive cycles in birds. *In* Giese, A. L. (ed), "Photophysiology" **6,** 325. Academic Press, N.Y. and London. [44]

Feller, W. (1957). "An Introduction to Probability Theory and Its Applications, 2nd edn". John Wiley, New York. [110]

Fisher, R. A. (1930). "The Genetical Theory of Natural Selection". Clarendon Press, Oxford. [145]

Fitzsimons, J. T. (1968). La soif extracellulaire. *Annls. Nutrit. aliment.* **22,** 131–144. [3]

Freeman, S. and McFarland, D. J. (1974). RATSEX: An exercise in simulation. *In* McFarland, D. J. (ed), "Motivational Control Systems Analysis", pp. 479–510. Academic Press, London. [81]

Freeman, S. and McFarland, D. J. (1981). "The Darwinian Objective Function". *In* McFarland, D. (ed.) "Functional Ontogeny". Pitman Books Ltd, London. (Unpublished.) [145–147], 160, 171]

Fry, F. E. J. and Hochachka, P. W. (1970). Fish. *In* Whittow, G. C. (ed.), "Comparative Physiology of Thermoregulation, vol. 1", pp. 79–134. Academic Press, N.Y. and London. [28]

Geertsema, S. and Reddingins, H. (1974). Preliminary considerations in the simulation of behaviour. *In* McFarland, D. J. (ed.), "Motivational Control Systems Analysis", pp. 355–405. Academic Press, London. [73]

Ghett, K. J. de (1978). Hierarchical cluster analysis. *In* Colgan, P. W. (ed.), "Quantitative Ethology", pp. 115–144. John Wiley & Sons, N.Y. [72]

Gilbert, E. G. (1963). Controllability and observability in multivariable control systems. *J. Soc. ind. appl. Math.* (Ser. A, Control) **1,** 128–151. [83]

Goss-Custard, J. D. (1977). Optimal foraging and size-selection of worms by redshank *Tringa totanus*. *Anim. Behav.* **25,** 10–29. [94, 95]

Gumma, N. R., South, F. E. and Allen, J. N. (1967). Temperature preference in golden hamsters. *Anim. Behav.* **15,** 534–537. [28]

Hailman, J. P. (1977). "Optical Signals". Indiana University Press, Bloomington and London. [38]

Halliday, T. R. (1975). An observational and experimental study of sexual behaviour in the smooth newt, *Triturus vulgaris* (Amphibia: Salamandridae). *Anim. Behav.* **23,** 291–322. [77, 78, 158, 160]

Halliday, T. R. (1976). The libidinous newt: An analysis of variations in the sexual behaviour of the male smooth newt, *Triturus vulgaris*. *Anim. Behav.* **24,** 398–414. [77]

Halliday, T. R. (1977). The effect of experimental manipulation of breathing behaviour on the sexual behaviour of the smooth newt, *Triturus vulgaris*. *Anim. Behav.* **25,** 39–45. [81]

Halliday, T. R. and Houston, A. I. (1978). The newt as an honest salesman. *Anim. Behav.* **26,** 1273–1274. [158]

Hauenschild, C. (1960). Lunar periodicity. *Cold Spring Harbour Symp. quant. Biol.* **25,** 491–497. [44]

Heiligenberg, W. (1966). The stimulation of territorial singing in the house cricket, (*Acheta domesticus*). *Z. vergl. Physiol.* **53,** 114–129. [41]

Heiligenberg, W. (1969). The effect of stimulus chirps on a cricket's chirping (*Acheta domesticus*). *Z. vergl. Physiol.* **65,** 70–97. [42]

Heiligenberg, W. (1974). A stochastic analysis of fish behaviour. *In* McFarland, D. J. (ed.), "Motivational Control Systems Analysis", pp. 87–118. Academic Press, London. [71, 77]

Heiligenberg, W. (1976). The interaction of stimulus patterns controlling aggressiveness in the cichlid fish *Haplochromis burtoni*. *Anim. Behav.* **24,** 452–458. [60]

Heiligenberg, W., Kramer, U. and Schulz, V. (1972). The angular orientation of the black eye-bar in *Haplochromis burtoni* (Cichlidae: Pisces) and its relevance to aggressivity. *Z. vergl. Physiol.* **76,** 168–176. [42]

Heller, R. and Milinski, M. (1979). Optimal foraging of sticklebacks on swarming prey. *Anim. Behav.* **27,** 1127–1141. [139, 140]

Herrnstein, R. J. (1958). Some factors influencing behaviour in a two-response situation. *Trans. N.Y. Acad. Sci.* **1,** 35–45. [101]

Herrnstein, R. J. (1970). On the law of effect. *J. exp. Analysis Behav.* **13,** 243–266. [61]

Heyman, G. M. (1979). Matching and maximising in concurrent schedules. *Psychol. Rev.* **86,** 496–500. [125]

Heyman, G. M. and Luce, R. D. (1979a). Operant matching is not a logical consequence of maximising reinforcement rate. *Animal Learning and Behaviour* **7,** 133–140. [126]

Heyman, G. M. and Luce, R. D. (1979b). Reply to Rachlin's comment. *Animal Learning and Behaviour* **7,** 269–270. [126]

Hinde, R. A. (1959). Unitary drives. *Anim. Behav.* **7,** 130–141. [5]

Hinde, R. A. (1970). "Animal Behaviour", 2nd. edn. McGraw-Hill, New York. [62]

Hinde, R. A. and Steel, F. A. (1966). Integration of the reproductive behaviour of female canaries. *Symp. soc. exp. Biol.* **20,** 401–426. [3]

Hogan, J. A., Kleist, S. and Hutchings, C. S. L. (1970). Display and food as reinforcers in the Siamese fighting fish (*Betta splendens*). *J. comp. Physiol. Psychol.* **70,** 351–357. [122, 123]

Hogan, J. and Roper, T. (1978). A comparison of the properties of different reinforcers. *In* Rosenblatt, J., Hinde, R., Beer, C. and Busnel, M-C. (eds.), "Advances in the Study of Behavior", vol. 8, pp. 155–225. Academic Press, New York. [122]

Holst, E. von and Saint Paul, V. von (1963). On the functional organisation of drives. *Anim. Behav.* **11,** 1–20. [3]

Houston, A. I. (1977). "Models of Animal Motivation". Unpublished D. Phil, thesis. University of Oxford. [60, 82, 83, 158]

Houston, A. I. (1980). Godzilla v. the creature of the black lagoon. *In* Toates, F. M. and Halliday, T. R. (eds), "The Analysis of Motivational Processes". Academic Press, London. [163, 180]

Houston, A. I., Halliday, T. R. and McFarland, D. J. (1977). Towards a model of the courtship of the smooth newt *Triturus vulgaris*, with special emphasis on problems of observability in the simulation of behaviour. *Med. Biol. Eng. Comput.* **15,** 49–61. [73, 78–80]

Houston, A. I. and McFarland, D. J. (1976). On the measurement of motivational variables. *Anim. Behav.* **24**, 459–475. [49, 56, 57]

Houston, A. I. and McFarland, D. J. (1980). Behavioural resilience and its relation to demand functions. *In* Staddon, J. E. R. (ed.), "Limits to Action: The Allocation of Individual Behavior". pp. 177–203. Academic Press, New York. [111, 112, 116, 165]

Houston, A. I. and McNamara, J. M. (1981). How to maximise reward rate on two variable interval paradigms. *J. exp. Analysis Behav.* **35**, 367–396. [126]

Huey, R. B. and Slatkin, M. (1976). Cost and benefit of lizard thermoregulation. *Q. Rev. Biol.* **51**, 364–384. [94]

Hutchinson, J. B. (1978). (ed.), "Biological Determinants of Sexual Behaviour". John Wiley and Sons, Chichester, U.K. [69]

Ishay, J., Bytinski-Salz, H. and Shulov, A. (1967). Contributions to the bionomics of the oriental hornet. (*Vespa orientalis* Fab.). *Israel J. Ent.* **2**, 45–106. [145]

Jacobs, O. L. R. (1967). "An Introduction to Dynamic Programming". Chapman and Hall, London. [132]

Jacobs, O. L. R. (1974). "Introduction to Control Theory". Clarendon Press, Oxford. [129, 136]

Kacelnik, A. (1979). "Studies of Foraging Behaviour in Great Tits (*Parus major*)". Unpublished D.Phil. thesis. University of Oxford. [184]

Kalman, R. E. (1963). Mathematical description of linear dynamical systems. *J.S.I.A.M. Control*, Series A, vol. **1**, 152–192. [83, 84, 86]

Kalman, R. E. (1968). New development in systems theory relevant to biology. *In* Mesarovic, M. C. (ed.), "Systems Theory and Biology", pp. 222–232. Springer-Verlag, Berlin. [84]

Kapur, P., Ghosh, P. and Nath, N. G. (1976). Stability, controlability and observability of arterial circulation. *J. theoret. Biol.* **61**, 15–19. [85]

Katz, P. L. (1974). A long-term approach to foraging optimization. *Am. Nat.* **108**, 758–782. [140]

Keeton, W. T. (1974). The orientational and navigational basis of homing in birds. *Advances in the Study of Behaviour* **5**, 48–132. [38, 41]

Kimble, G. A. (1961). Hilgard and Marguis "Conditioning and Learning", 2nd edn. Appleton-Century Crofts, N.Y. [176]

King, B. M. and Gaston, M. G. (1976). Factors influencing the hunger and thirst motivated behavior of hypothalamic hyperphagic rats. *Physiol. Behav.* **16**, 33–41. [124]

Klein, H. (1974). The adaptational value of internal annual clocks in birds. *In* Pengelley, E. T. (ed.), "Circannual Clocks". Academic Press, N.Y. [44]

Klitzner, M. D. and Anderson, N. H. (1977). Motivation × Expectancy × Value: A functional measurement approach. *Motivation and Emotion* **1,** 347–365. [60]

Krantz, D. H., Luce, R. D., Suppes, P. and Tversky, A. (1971). "Foundations of Measurement, vol. **1,** Additive and Polynomial Representations". Academic Press, New York. [49]

Krantz, D. H. and Tversky, A. (1971). Conjoint-measurement analysis of composition rules in psychology. *Psychol. Rev.* **78,** 151–169. [53, 54]

Krebs, J. R. (1973). Behavioural aspects of predation. *In* Bateson, P. P. E. and Klopfer, P. H. (eds.), "Perspectives in Ethology", pp. 73–111. Plenum Press, New York. [95]

Krebs, J. R. (1978). Optimal foraging: decision rules for predators. *In* Krebs, J. R. and Davies, N. B. (eds.), "Behavioural Ecology", pp. 23–63. Blackwell Scientific Publications, Oxford. [94, 95, 135]

Krebs, J. R., Erichson, J. T., Webber, M. I. and Charnov, E. L. (1977). Optimal prey selection in the great tit (*Parus major*). *Anim. Behav.* **25,** 30–38. [140]

Krebs, J. R., Kacelnik, A. and Taylor, P. (1978). Test of optimal sampling by foraging great tits. *Nature, Lond.* **275,** 27–31. [183]

Lack, D. (1943). "The Life of the Robin". Cambridge University Press. [38]

Lack, D. (1966). "Population Studies of Birds". Clarendon Press, Oxford. [180]

Lancaster, K. (1968). "Mathematical Economics". Macmillan, London. [150]

Larkin, S. (1981). "Time and energy in decision-making". Unpublished D.Phil. thesis. University of Oxford. [143]

Larkin, S. and McFarland, D. (1978). The cost of changing from one activity to another. *Anim. Behav.* **26,** 1237–1246. [99, 142]

Lea, S. E. G. (1978). The psychology and economics of demand. *Psychol. Bull.* **85,** 441–466. [119]

Lea, S. E. G. and Roper, T. J. (1977). Demand for food on fixed-ratio schedules as a function of the quality of concurrently available reinforcement. *J. exp. Analysis Behav.* **27,** 371–380. [100]

Lehrman, D. S. (1959). Hormonal responses to external stimulation in birds. *Ibis* **101,** 478–496. [3]

Leong, C. Y. (1969). The quantitative effect of releasers on the attack readiness of the fish *Haplochromis burtoni* (Cichlidae: Pisces). *Z. vergl. Physiol.* **65,** 29–50. [42]

Lloyd, I. H. (1974). Stochastic identification methods. *In* McFarland, D. J. (ed.), "Motivational Control Systems Analysis", pp. 169–207. Academic Press, London. [72]

Lloyd, I. H. (1975). "The Application of Systems Analysis Techniques to the Interaction of Hunger and Thirst in the Dove". Unpublished Ph.D. thesis. University of London. [72]

Lorenz, K. (1965). "Evolution and Modification of Behaviour". Methuen, London. [169]

Luce, R. D. and Tukey, J. W. (1964). Simultaneous conjoint measurement: a new type of fundamental measurement. *J. math. Psychol.* **1,** 1–27. [58]

MacArthur, R. H. and Pianka, E. R. (1966). On the optimal use of a patchy environment. *Am. Nat.* **100,** 603–609. [95]

McCleery, R. H. (1977). On satiation curves. *Anim. Behav.* **25,** 1005–1015. [72, 137, 154]

McCleery, R. H. (1978). Optimal behaviour sequences and decision making. In Krebs, J. R. and Davies, N. B. (eds.), "Behavioural Ecology", pp. 377–410. Blackwell Scientific Publications, Oxford. [93, 153, 156]

Macevicz, S. and Oster, G. (1976). Modelling social insect populations. II. Optimal reproductive strategies in annual eusocial insect colonies. *Behav. Ecol. Sociobiol.* **1,** 265–282. [144, 145]

McFarland, D. J. (1964). Interaction of hunger and thirst in the Barbary dove. *J. comp. physiol. Psychol.* **58,** 174–179. [3, 22]

McFarland, D. J. (1965a). The effect of hunger on thirst motivated behaviour in the Barbary dove. *Anim. Behav.* **13,** 286–292. [3]

McFarland, D. J. (1965b). Control theory applied to the control of drinking in the Barbary dove. *Anim. Behav.* **13,** 478–492. [71, 74]

McFarland, D. J. (1966). On the causal and functional significance of displacement activities. *Z. Tierpsychol.* **23,** 217–235. [44]

McFarland, D. J. and McFarland, F. J. (1968). Dynamic analysis of an avian drinking response. *Med. Biol. Engng.,* **6,** 659–668. [163]

McFarland, D. J. (1970). Behavioural aspects of homeostasis. In "Advances in the Study of Behaviour", vol. 3, pp. 1–26. Academic Press, New York. [51, 69, 73, 74]

McFarland, D. J. (1971). "Feedback Mechanisms in Animal Behaviour". Academic Press, London and New York.
[1, 2, 11, 19, 22, 50, 51, 70–76, 84, 163]

McFarland, D. J. (1973). Stimulus relevance and homeostasis. In Hinde, R. A. and Stevenson-Hinde, J. (eds.), "Constraints on Learning", pp.141–155. Academic Press, London. [29, 38, 69, 163, 174]

McFarland, D. J. (1974a). Experimental investigation of motivational state. In McFarland, D. J. (ed.), "Motivational Control Systems Analysis", pp. 251–282. Academic Press, London and New York. [74]

McFarland, D. J. (1974b). Time-sharing as a behavioural phenomenon. In Lehrman, D. S., Rosenblatt, J. S., Hinde, R. A. and Shaw, E. (eds.), "Advances in the Study of Behavior", vol. 5, pp. 201–225. Academic Press, New York. [89]

McFarland, D. J. (1976). Form and function in the temporal organisation of behaviour. *In* Bateson, P. P. G. and Hinde, R. A. (eds.), "Growing Points in Ethology", pp. 55–93. Cambridge University Press. [103, 134]

McFarland, D. J. (1977). Decision-making in animals. *Nature, Lond.* **269,** 15–21. [45, 79, 92, 93, 153, 156, 160]

McFarland, D. J. (1978a). Hunger in interaction with other aspects of motivation. *In* Booth, D. A. (ed.), "Hunger Models: Computable Theory of Feeding Control", pp. 375–405. Academic Press, London and New York. [32, 70, 74, 170]

McFarland, D. J. (1978b). Optimality considerations in animal behaviour. *In* Reynolds, V. and Blurton-Jones, N. (eds.), "Human Behaviour and Adaptation", pp. 53–76. Taylor & Francis Ltd, London. [172–174]

McFarland, D. J. and Baher, E. (1968). Factors affecting feather posture in the Barbary dove. *Anim. Behav.* **16,** 171–177. [58, 59]

McFarland, D. J. and Budgell, P. (1970). The thermoregulatory role of feather movements in the Barbary dove (*Streptopelia risoria*). *Physiol. Behav.* **5,** 763–771. [3, 58, 59, 73]

McFarland, D. J. and Nunez, A. T. (1978). Systems analysis and sexual behaviour. *In* Hutchison, J. B. (ed.), "Biological Determinants of Sexual Behaviour", pp. 615–652. John Wiley, Chichester. [28, 73, 74]

McFarland, D. J. and Rolls, B. J. (1972). Suppression of feeding by intracranial injections of angiotensin. *Nature, Lond.* **236,** 172–173. [5]

McFarland, D. J. and Sibly, R. M. (1972). 'Unitary drives' revisited. *Anim. Behav.* **20,** 548–563. [5, 6, 30, 32, 33, 84, 87, 88]

McFarland, D. J. and Sibly, R. M. (1975). The behavioural final common path. *Phil. Trans. R. Soc.* (Series B) **270,** 265–293.
[3, 5, 8–12, 16, 39, 48, 106]

McFarland, D. J. and Wright, P. (1969). Water conservation by inhibition of food intake. *Physiol. & Behav.* **4,** 95–99. [22]

Mackintosh, N. J. (1973). Stimulus selection: Learning to ignore stimuli that predict no change in reinforcement. *In* Hinde, R. A. and Stevenson-Hinde, J. (eds.), "Constraints on Learning", pp. 75–100. Academic Press, London. [38, 174]

Mackintosh, N. J. (1974). "The Psychology of Animal Learning". Academic Press, London. [169, 184]

McNamara, J. and Houston, A. I. (1980). The application of statistical decision theory to animal behaviour. *J. theoret. Biol.* **85,** 673–690. [180, 181, 184]

Matthews, G. V. T. (1965). "Bird Navigation". Cambridge University Press, London and N.Y. [47]

Maynard Smith, J. (1978). Optimization theory in evolution. *A. Rev. Ecol. Syst.* **9,** 31–56. [92]

Meddis, R. (1975). On the function of sleep. *Anim. Behav.* **23,** 676–691. [46]

Metz, H. A. J. (1974). Stochastic models for the temporal fine structure of behavioural sequences. *In* McFarland, D. J. (ed.), "Motivational Control Systems Analysis", pp. 5–86. Academic Press, London. [72]

Metz, H. A. J. (1977). State space models for animal behaviour. *Annals of Systems Research* **6**, 65–109. [87]

Milinski, M. and Heller, R. (1978). Influence of a predator on the optimal foraging behaviour of sticklebacks (*Gasterosteus aculeatus* L.). *Nature, Lond.* **275**, 642–644. [137, 139]

Miller, H. L. Jr. (1976). Matching based hedonic scaling in the pigeon. *J. exp. Analysis Behav.* **26**, 335–347. [61, 62]

Milsum, J. H. (1966). "Biological Control Systems Analysis". McGraw-Hill, New York. [19]

Minsky, M. (ed.) (1968). "Semantic Information Processing". MIT Press, Cambridge, Mass. [45]

Mittelstaedt, M. (1964). Basic control patterns of orientational homeostasis. *Symp. Soc. exp. Biol.* **18**, 365–385. [170]

Mogenson, G. J. and Calaresu, F. R. (1978). Food intake considered from the viewpoint of systems analysis. *In* Booth, D. A. (ed.), "Hunger Models", pp. 1–24. Academic Press, London. [69]

Moore, E. F. (1956). Gedanken-experiments on sequential machines. *In* Shannon, C. E. and McCarthy, J. (eds.), "Automata Studies", pp. 129–153. Princeton University Press. [86]

Morris, D. (1957). 'Typical intensity' and its relation to the problem of ritualisation. *Behaviour* **11**, 1–12. [60]

Mrosovsky, N. (1968). The adjustable brain of hibernators. *Scient. Amer.* **218** (3), 110–116. [122–124]

Navarick, D. J. and Fantino, E. (1974). Stochastic transitivity and unidimensional behavior theories. *Psychol. Rev.* **81**, 426–441. [9, 106]

Neill, S. R. St. J. and Cullen, J. M. (1974). Experiments on whether schooling by their prey affects the hunting behaviour of cephalopods and fish predators. *J. Zool., Lond.* **172**, 549–569. [138]

Nelson, K. (1965). The temporal patterning of courtship behaviour in the glandulocaudine fishes (*Ostanophysi, Characidae*). *Behaviour* **24**, 90–146. [87]

Nelson, K. (1965). After-effects of courtship in the male three-spined stickleback. *Z. vergl. Physiol.* **50**, 569–597. [87]

Ng, Y. (1977). Sub-semiorder: A model of multidimensional choice with preference intransitivity. *J. math. Psychol.* **16**, 51–59. [9]

Norman, J. M. (1975). "Elementary Dynamic Programming". Edward Arnold, London. [129]

Norman, M. F. and Gallistel, C. R. (1978). What can one learn from a strength-duration experiment? *J. math. Psychol.* **18**, 1–24. [86]

Oatley, K. (1967). A control model for the physiological basis of thirst. *Med. Biol. Engng.*, **5**, 225–237. [71, 74, 86]

Oatley, K. (1971). Dissociation of the circadian drinking pattern from eating. *Nature, Lond.* **229,** 494–496. [47]

Oatley, K. (1974). Circadian rhythms in motivational systems. *In* McFarland, D. J. (ed.), "Motivational Control Systems Analysis", pp. 427–459. Academic Press, London. [45, 47]

Oatley, K. and Toates, F. M. (1971). Frequency analysis of the thirst control system. *Nature, Lond.* **232,** 562–564. [73]

Oatley, K. and Tonge, D. A. (1969). The effect of hunger on water intake in rats. *Q. Jl exp. Psychol.* **21,** 162–171. [3]

Oster, G. F. and Wilson, E. O. (1978). "Caste and Ecology in the Social Insects". Princeton University Press. [92, 145]

Pengelley, E. T. (1974). "Circannual Clocks". Academic Press, New York. [171]

Perrins, C. M. (1979). "British Tits". Collins, London. [140]

Pfanzagl, J. (1968). "Theory of Measurement". Physica-Verlag, Würzburg. [49, 60]

Pimental, R. A. and Frey, D. F. (1978). Multivariate analysis of variance and discriminant analysis. *In* Colgan, P. W. (ed.), "Quantitative Ethology". John Wiley and Sons, New York. [72]

Pring-Mill, F. (1979). Tolerable feedback; a mechanism for behavioural change. *Anim. Behav.* **27,** 226–236. [44]

Prosser, C. L. (1958). The nature of physiological adaptation. *In* Prosser, C. L. (ed.), "Physiological Adaptation". Am. Physiol. Soc., Washington DC. [20]

Prosser, C. L. (1973). "Comparative Animal Physiology", 3rd edn. W. B. Saunders, Philadelphia, London, Toronto. [20]

Pulliam, H. R. (1974). On the theory of optimal diets. *Am. Nat.* **108,** 59–74. [95]

Rachlin, H. (1978). A molar theory of reinforcement schedules. *J. exp. Analysis Behav.* **30,** 345–360. [124, 125]

Rachlin, H. (1979). Comment on Heyman and Luce: "Operant matching is not a logical consequence of maximising reinforcement rate". *Animal Learning and Behaviour,* **7,** 267–268. [126]

Rao, S. S. (1978). "Optimization – Theory and Applications". Wiley Eastern Ltd., New Delhi. [129]

Revusky, S. and Garcia, J. (1970). Learned aversions over long delays. *In* Bower, G. H. (ed.), "Psychology of Learning and Motivation" **4,** 1–83. Academic Press, New York. [69, 169, 174]

Richards, S. A. (1975). Thermal homeostasis in birds. *In* Peaker, M. (ed.), "Avian Physiology". Academic Press, London. [37]

Richter, C. P. (1943). Total self-regulatory functions in animals and human beings. *Harvey Lectures* **38,** 63–103. [30]

Riggs, D. S. (1970). "Control Theory and Physiological Feedback Mechanisms". The Williams & Wilkins Co., Baltimore. [22]

Rilling, S., Mittelstaedt, H. and Roeder, K. O. (1959). Prey recognition in the praying mantis. *Behaviour* **14**, 164–184. [41]

Rolls, B. J. and McFarland, D. J. (1973). Hydration releases inhibition of feeding produced by intracranial angiotensin. *Physiology Behav.* **11**, 881–884. [3]

Rosen, R. (1967). "Optimality Principles in Biology". Butterworths, London. [154]

Rosen, R. (1970). "Dynamical Systems Theory in Biology", vol. **1**. John Wiley, New York. [19, 82, 149]

Rozin, P. and Kalat, J. W. (1971). Specific hungers and poison avoidance as adaptive specializations in learning. *Psychol. Rev.* **78**, 459–486.
[29, 38, 69, 163, 169, 174]

Rozin, P. and Mayer, J. (1961). Thermal reinforcement and thermoregulatory behaviour in the goldfish. *Science, N.Y.* **134**, 942–943. [28]

Russek, M. (1978). Semi-quantitative simulation of food intake control and weight regulation. *In* Booth, D. A. (ed.), "Hunger Models", pp. 195–226. Academic Press, London. [70]

Schneider, D. (1969). Insect olfaction: Deciphering system for chemical messages. *Science, N.Y.* **163**, 1031–1036. [44]

Schoener, T. W. (1971). Theory of feeding strategies. *A. Rev. Ecol. Syst.* **2**, 369–404. [95]

Schultz, D. G. and Melsa, J. L. (1967). "State Functions and Linear Control Systems". McGraw-Hill, New York. [75, 76, 83, 84, 134]

Seitz, A. (1940). Die Paarbildung bei einigen Cichliden. *Z. Tierpsychol.* **4**, 40–84. [41]

Seligman, M. E. P. (1970). On the generality of the laws of learning. *Psychol. Rev.* **77**, 406–418. [169]

Sherry, D. F., Mrosovsky, N. and Hogan, J. A. (1980). Weight loss and anorexia during incubation in birds. *J. comp. physiol. Psychol.* **94**, 89–98. [165]

Sibly, R. M. and McFarland, D. J. (1974). A state-space approach to motivation. *In* McFarland, D. J. (ed.), "Motivational Control Systems Analysis", pp. 213–250. Academic Press, London and New York. [9, 20–34]

Sibly, R. M. and McFarland, D. J. (1976). On the fitness of behaviour sequences. *Am. Nat.* **110**, 610–617. [91, 112, 135, 143, 146]

Staddon, J. E. R. (1976). Learning as adaptation. *In* Estes, W. K. (ed.), "Handbook of Learning and Cognitive Processes", vol. 2, pp. 37–98. Erlbaum Associates, New York. [169]

Staddon, J. E. R. (1977). On Herrnstein's equation and related forms. *J. exp. Analysis Behav.* **28**, 163–170. [124]

Staddon, J. E. R. (1979). Operant behaviour as adaptation to constraint. *J. exp. Psychol.* **108,** 48–67. [124]

Staddon, J. E. R. and Motheral, S. (1978). On matching and maximising in operant choice experiments. *Psychol. Rev,* **85,** 436–444. [125]

Staddon, J. E. R. and Motheral, S. (1979). Response independence, matching and maximising: A reply to Heyman. *Psychol. Rev.* **86,** 501–505. [126]

Teitelbaum, P. (1961). Disturbances in feeding and drinking behaviour after hypothalamic lesions. *In* "Nebraska Symposium on Motivation, 1961". University of Nebraska Press. [122, 124]

Thorndike, E. L. (1911). "Animal Intelligence: experimental studies". Macmillan, New York. [169]

Tinbergen, N. (1951). "The Study of Instinct". Clarendon Press, Oxford. [169]

Toates, F. M. (1975). "Control Theory in Biology and Experimental Psychology". Hutchinson Educational, London. [70, 73, 74]

Toates, F. M. (1979). Homeostasis and drinking. *The Behavioural and Brain Sciences* **2,** 95–139. [74]

Toates, F. M. and Archer, J. A. (1978). A comparative review of motivation systems using classical control theory. *Anim. Behav.* **26,** 368–380. [81]

Toates, F. M. and O'Rourke, C. (1978). Computer simulation of male rat sexual behaviour. *Med. Biol. Eng. Comput.* **16,** 98–104. [81]

Tovish, A. (1981). Learning to alter the availability and accessibility of resources. *In* McFarland, D. (ed.) "Functional Ontogeny". Pitman Books Ltd, London. (Unpublished.) [174, 178, 179]

Trivers, R. L. (1972). Parental investment and sexual selection. *In* Campbell, B. (ed.), "Sexual Selection and the Descent of Man". Aldine, Chicago. [131]

Tversky, A. (1969). Intransitivity of preferences. *Psychol. Rev.* **76,** 31–48. [9]

Wells, G. P. (1966). The lugworm (*Arenicola*): a study in adaptation. *Netherlands J. Sea Res.* **3,** 294–313. [106]

Ward, P. (1965). Feeding ecology of the black-faced *Quelea quelea* in Nigeria. *Ibis* **107,** 173–214. [141]

Werner, E. E. and Hall, D. J. (1974). Optimal foraging and the size selection of prey by the Bluegill sunfish (*Lepomis macrochirus*). *Ecology* **55,** 1042–1052. [94, 95]

Wilson, E. O. (1975). "Sociobiology". Harvard University Press, Cambridge, Mass. [144]

Wooton, R. J. (1971). A note on the nest-raiding behaviour of male sticklebacks. *Can. J. Zool.* **49,** 960–962. [89]

Zeigler, H. P. (1974). Feeding behaviour in the pigeon: A neurobehavioral analysis. *In* Goodman, I. and Schein, M. (eds.), "Birds: Brain and Behavior", pp. 101–132. Academic Press, New York. [11]

Zepelin, H. and Rechtschaffen, A. (1974). Mammalian sleep, longevity, and energy metabolism. *Brain, Behav. Evol.* **10,** 425–470. [46]

Index

Acclimatization, 21, 22, 23, 25, 26, 28, 29, 30, 164–6, 170
Action, 7, 11, 15, 16
Activity, 7, 16
Adaptation
 space, 23, 35
 theorem, 23–5, 35
Additive representation, 56
Ambivalent environment, 34, 36

Bandit problem, 180, 183–4
Bang-bang control, 145
Barbary dove, 22, 99, 163
Bayes' rule, 181–2
Behaviour space, 26, 35
Behavioural repertoire, 10, 16
Burmese red jungle fowl, 165

Candidate
 space, 8, 10–11, 14, 15, 16
 state, 8, 9, 11, 16
Causal
 factor space, 6, 7, 10, 16, 31, 32
 factor state, 6, 16, 39
Cichlid, 71, 77
Combination rule, 49
Command
 space, 29, 34, 35
  ideal, 29, 36
  minimal, 29, 30, 36
  realizable, 35, 36
 state, 32, 34, 51
Concorde fallacy, 131
Concurrence, 53, 54–5.
Cone of possible consequences, 34, 36
Confusion effect, 138
Conjoint measurement, 52 *et seq.*
Consequence space, 34
Controllable system, 83, 84
Cost, 103, 108
Cost function, 93, 107, 108, 153–5, 161–2, 164, 166, 171, 173
Costate variables, 133–4, 149–51

Cue
 space, 39, 48
 state, 39, 48
 strength, 39, 48

Decision rules, 105, 134, 160 *et seq.*, 170, 172
Digger wasp, 161
Dimensionality theorem, 30, 36
Displacement space, 26, 35
Double cancellation, 53, 55–6, 66
Drift, 25
Dynamic programming, 130–32, 141–4, 183

Elasticity of demand, 100–102, 108, 119 *et seq.*

Final common path, 3, 5, 11
Finite automaton, 86
Fitness, 90, 107, 108, 110, 146, 169, 171, 172
Functional measurement, 58 *et seq.*

Goal function, 104–107, 108, 153–5, 164, 165, 166, 171, 173
Golden hamster, 27
Great tit, 140
Guppy, 13, 60

Herring gull, 39, 45
Homeostasis, 1, 22, 29, 73–5

Interval scale, 52, 64
Inverse optimality, 106, 153 *et seq.*

Lagrange multiplier, 113, 117–19, 133, 150–51
Lethal boundary, 20, 35, 110, 111, 126, 135
Linear fan, 59
Loggerhead shrike, 141

Matching law, 60, 61, 124–6
Minimal system, 84
Motivational isocline, 12–15, 17, 63

Nerode equivalence, 86, 87

203

Objective function, 92, 104–107, 108, 129, 147, 148, 153–5, 162
Observability, 83 et seq.
Optimal foraging theory, 95, 135
Ordinal scale, 50, 52, 63

Phase variables, 76
Physiological
  space, 20, 28, 35
  state, 20, 21, 22, 23, 29, 32, 35
Pied flycatcher, 42
Pig, 37
Pigeon, 11, 37, 61
Pigmy owl, 42
Plant equations, 133, 169, 173
Pontryagin's maximum principle, 132 et seq., 148–9, 151
Posterior distribution, 181–3
Principle of optimality, 130
Principle of superposition, 85
Prior distribution, 180–81

Rat, 37, 47, 62, 81, 122
Red-backed shrike, 42

Redshank, 95, 96, 98
Regulatory space, 26, 35
Resilience, 112, 119 et seq
Robin, 38

Satiation curve, 72, 136–7, 154–5
Siamese fighting fish, 122
Smooth newt, 77, 155
Sparrowhawk, 140
Stickleback, 87, 89
Switching line, 9, 10, 15, 17, 144

Tendency, 8, 11, 14, 15, 16, 49, 54 et seq., 103
Trajectory, 14, 85, 105, 135
Transitivity, 9, 65, 106

Univalent environment, 33, 36
Utility, 91, 108
Utility function, 92, 108

Weaver bird, 140
White-crowned sparrow, 44